Collins

IB Science Skills
Chemistry

Chris Conoley

William Collins' dream of knowledge for all began with the publication of his first book in 1819. A self-educated mill worker, he not only enriched millions of lives, but also founded a flourishing publishing house. Today, staying true to this spirit, Collins books are packed with inspiration, innovation and practical expertise. They place you at the centre of a world of possibility and give you exactly what you need to explore it.

Collins. Freedom to teach

Published by Collins
An imprint of HarperCollinsPublishers
77–85 Fulham Palace Road
Hammersmith
London
W6 8JB

Browse the complete Collins catalogue at
www.collins.co.uk

© HarperCollins*Publishers* Limited 2014

10 9 8 7 6 5 4 3 2 1

ISBN-13 978 0 00 755468 3

British Library Cataloguing in Publication Data
A Catalogue record for this publication is available from the British Library

Written by **Chris Conoley**

Commissioned by
Lucy Killick

Project managed, edited and proofread by
Cambridge Editorial

Indexed by **Chris Bell**

Production by **Emma Roberts**

Typeset by **Jouve India Private Limited**

New illustrations by **Ann Paganuzzi**

Picture research by **Amanda Redstone**

Interior design by **Anna Plucinska**

Cover design by **Angela English**

With thanks to our reviewers: **Mark Levesley, Chris Curtis, Jon Tootill** and **Maurice Carmody**

Printed and bound by **L.E.G.O. S.p.A. Italy**

Acknowledgements

The publishers wish to thank the following for permission to reproduce photographs. Every effort has been made to trace copyright holders and to obtain their permission for the use of copyright material. The publishers will gladly receive any information enabling them to rectify any error or omission at the first opportunity.

Cover photo Yellowj/Shutterstock

Fig 1 Eye Ubiquitous/Superstock, Fig 2 petarg/Shutterstock, Fig 3 RGtimeline/Shutterstock, Fig 6a Lindsey Moore/Shutterstock, Fig 9 Dmitry Kalinovsky/Shutterstock, Fig 18 wavebreakmedia/Shutterstock, Fig 37 jupeart/Shutterstock, Fig 38 mathagraphics/Shutterstock, Fig 45 Neil Bartlett, Chemistry, UBC 41.1/1944-1, B.C. Jennings/University of British Columbia Archives, Fig 46 Sebastian Duda/Shutterstock, Fig 48 Againstar/Shutterstock, Fig 49 Alexandru Nika/Shutterstock, Fig 50 Tyler Boyes/Shutterstock, Fig 52 ogwen/Shutterstock, Fig 53 Johnson Matthey PLC, Fig 58 Dmitri Melnik/Shutterstock, Fig 60 Volosina/Shutterstock, Fig 65 ZRyzner/Shutterstock, Fig 66 Mikhail Zahranichny/Shutterstock, Fig 69 Hung Chung Chih/Shutterstock.

Contents

Getting the best from the book 1

SKILLS

Working Scientifically

S1 Explaining observations: hypotheses; predictions; theories; models 4

S2 Understanding how science advances: using existing knowledge; posing scientific questions..6

S3 Using appropriate techniques: using valid procedures; using IT; using mass spectrometry; combining mass spectrometry and gas chromatography........5

S4 Making valid observations: and what makes an observation valid...............................6

S5 Taking measurements: accurate measurements; precise measurements 7

S6 Managing risk in investigations: identifying hazards; assessing risk...........................7

S7 Recording results appropriately: qualitative data; quantitative data; devising tables; tables to record titrations...............................8

S8 Analysing and interpreting data: trends and patterns; correlations; causal links 9

S9 Identifying anomalous data: and what to do with it 10

S10 Evaluating methodology, data and evidence: reliable data; valid data; accurate data 10

S11 Assessing error: percentage errors; human errors; random errors; systematic errors.....11

S12 Drawing valid conclusions: what conclusions are; and what makes them valid..12

S13 Communicating scientific information: scientific papers; scientific conferences; peer review; bias..................................12

S14 Considering ethical issues: what ethics means; what to consider 13

S15 Science and decision-making: benefits, drawbacks and risks for society 13

S16 Writing balanced equations: molecular formulae..14

S17 Relative atomic mass: definition; isotopes; mass number; mass spectrometers and mass spectra 15

S18 Amounts and the mole: amount and the mole unit; worked examples of mole calculations...15

S19 Relative molecular masses: calculating relative molecular masses.........................16

S20 Calculating amounts and masses from chemical equations: amounts from chemical equations; theoretical yield; worked examples..................................17

S21 Molar gas volumes: Avogadro's hypothesis 18

S22 Calculating volumes of gases from chemical equations: predicting volume of gas produced in reactions; worked examples 19

S23 Working with amounts in aqueous solution: calculating concentrations; worked examples..................................20

S24 Titrations: worked example of a titration calculation 21

Quality of Written Communication

S25 Writing for your intended audience: questions to ask yourself.........................22

S26 Ensuring meaning is clear: legibility, grammar, punctuation and spelling............22

S27 Organising information clearly and coherently: planning; organisation of answers; presentation; clarity 23

S28 Using specialist vocabulary: correct chemical terminology 23

Maths

S29 Standard form: what it is; adding and subtracting numbers in standard form; multiplying and dividing numbers in standard form.......................................24

S30 Ratios, fractions and percentages: worked examples of isotopic composition and percentage by mass calculations 26

S31 Using your calculator: standard form calculations; logarithms and antilogarithms; estimating 27

S32 Using an appropriate number of significant figures: determining the number of significant figures; decimal places in calculations; significant figures in calculations; rounding; calculations with more than one step 29

S33 Finding arithmetic means: means in volumetric analysis 30

S34 Changing the subject of an equation: worked examples 30

S35 Using logarithms: making numbers more manageable and patterns easier to see; ionisation energies; the pH scale 32

S36 Plotting graphs: worked examples using rates of reaction; graph plotting tips; discrete and continuous data 32

S37 Determining slopes and working out rates of reaction: direct proportionality; finding tangents; initial rates of reaction 34

S38 Orders of reaction from graphs: the rate equation; zero, first and second order reactions 35

S39 Appreciating, visualising and representing molecules in 2D and 3D: displayed formulae 36

Skills to Activities table 37

ACTIVITIES

A1 Iron-60 found on Earth 39
A2 The nuclear model of the atom 40
A3 The arrangement of electrons in shells 41
A4 The arrangement of electrons in subshells 42
A5 Noble gases react! 43
A6 The search for new medicines 44
A7 Bioethanol and Brazil 45
A8 Hydrogen: car fuel of the future? 46
A9 Fullerenes – new forms of carbon 47

A10 Manufacturing nitric acid – a greener way 48
A11 A structural model for benzene 50
A12 TNT – a formidable explosive 51
A13 Peppermint in medicine 52
A14 Saturated and unsaturated fatty acids 53
A15 Aramids: fire-resistant and bulletproof 55
A16 Reaction kinetics and vehicle exhausts 56

ASSESSING INVESTIGATIVE SKILLS

AIS1 58
AIS2 61

ANSWERS

A1 Iron-60 found on Earth 64
A2 The nuclear model of the atom 65
A3 The arrangement of electrons in shells 65
A4 The arrangement of electrons in subshells 66
A5 Noble gases react! 67
A6 The search for new medicines 67
A7 Bioethanol and Brazil 69
A8 Hydrogen: car fuel of the future? 69
A9 Fullerenes – new forms of carbon 70
A10 Manufacturing nitric acid – a greener way 71
A11 A structural model for benzene 73
A12 TNT – a formidable explosive 74
A13 Peppermint in medicine 75
A14 Saturated and unsaturated fatty acids 77
A15 Aramids: fire-resistant and bulletproof 78
A16 Reaction kinetics and vehicle exhausts 79
AIS1 80
AIS2 81
QWC Worked Examples 84

Glossary 87
Periodic table 89
Index 90

Getting the best from the book

We have designed this book to give you all the support you need to master the key skills necessary for success on your course. The science, maths and quality of written communication skills (QWC), for every major exam specification, are explained in detail. Whether you want more help with calculations using the mole, using the correct terminology or changing the subject of an equation, they are all explained in the different skills sections: Working Scientifically, Quality of Written Communication and Maths.

There are activities to practise each and every skill so you have a chance to apply your learning. They are set in interesting chemical contexts and give you the chance to apply your skills to unfamiliar situations. We have included the answers so you can check your understanding and also see how you can improve using the helpful hints, tips and pointers. The Assessing Investigative Skills activities give you the opportunity to practise some of the skills you will require in assessed practical tasks. To help you improve the quality of your written communication, we have included worked examples to show how low, medium and high answers get their marks. There is also a glossary to check your understanding of key terms.

SKILLS

Each skill is colour-coded to tell you what type of skill it is:
Working Scientifically
Quality of Written Communication
Maths.

Each skill is explained in detail.

The activities that practise the skill are listed at the bottom.

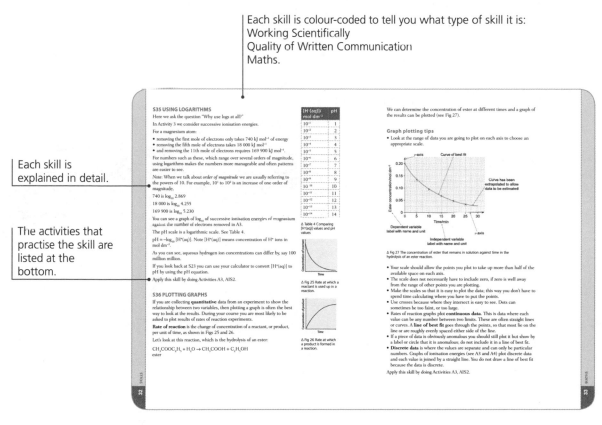

ACTIVITIES

The activities are listed in a logical sequence to allow you to tackle increasingly complex ideas as your knowledge of chemistry deepens.

Each activity includes a QWC question that asks you to think about your quality of written communication.

The skills practised in the questions are listed at the bottom of each activity.

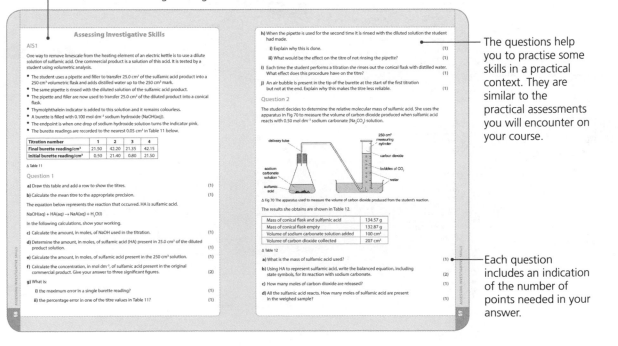

A8 HYDROGEN: CAR FUEL OF THE FUTURE?

One major disadvantage of using hydrogen as a car fuel is that it is difficult to store. One way would be to liquefy it but this requires energy and even then hydrogen has to be kept below −253 °C. Another possible method is to store it in a solid form – and this does not mean freezing it! Metal hydrides are compounds containing hydrogen; one candidate is magnesium hydride MgH_2. A litre of magnesium hydride contains almost as much hydrogen as a litre of liquefied hydrogen, although it is much heavier.

The first carbon nanotube was prepared in Japan by Iijima Sumio in 1991, spurred on by the discovery of the first fullerene (see A9). This single-walled tube is like a layer of graphite wound into a cylinder with a diameter ranging from 1 to 50 nm. Many uses have been suggested for nanotubes, one of which is for the on-board storage of hydrogen, where hydrogen is adsorbed onto the nanotube walls at only slightly increased pressure.

At first the uptake of hydrogen by nanotubes was only modest, but then researchers in Germany began to design a material based on the network found in natural sponges. They used advanced mathematics and computer models to suggest that the uptake could be greatly increased using a network of nanotubes. The combined weight of the nanotubes and the hydrogen adsorbed is predicted to be 5.5% of the total weight of the vehicle and the material would be cheap, non-toxic and lightweight.

Several groups of researchers are trying to make these sponge-like networks of nanotubes but, to date, no one has succeeded. The German team are also setting about making the material experimentally.

△ Fig 50 A single nanotube.

QUESTIONS

1. What is the percentage by mass of hydrogen in magnesium hydride? (A_r: Mg = 24.3; H = 1.0)
2. Hydrogen is released by the reaction of magnesium hydride with water to produce magnesium hydroxide.
   a) Construct an equation for this reaction and include state symbols.
   b) Calculate the volume of hydrogen produced per kilogram of magnesium hydride at 298 K and 101 kPa.
3. When hydrogen is adsorbed, what intermolecular force is likely to hold the hydrogen on the walls of the nanotubes? Explain your reasoning.
4. Hydrogen is easily released from the nanotube matrix inside the fuel tank. As the engine takes in hydrogen, the pressure in the tank falls, which releases more hydrogen. Use your understanding of dynamic equilibria to explain why more hydrogen is released.
5. What are the advantages and disadvantages of switching from petrol to hydrogen as a fuel? [QWC]

Skills practised

1, 3, 14, 15, 16, 18, 19, 20, 21, 22, 25, 26, 27, 28, 29, 30, 31, 32

A9 FULLERENES – NEW FORMS OF CARBON

In the 1970s Harry Kroto, a chemist from Sussex University, together with two Canadian astronomers, detected long chains of carbon atoms in interstellar space. Harry's hypothesis was that they might have formed in the carbon-rich clouds of red giant stars. He had the opportunity to test his idea when he met two American scientists, who were working with a laser machine that fired pulses reaching temperatures far higher than in most stars. In 1985, they vaporised graphite in an atmosphere of helium and analysed the resulting carbon species using a mass spectrometer.

The experiment did find evidence for the formation of long-chain carbon molecules but it also produced a very strange result: a large peak at 720 m/e on the mass spectrum. This corresponded to a C_{60} molecule. They worked out that the structure for this species was similar to the geodesic domes designed by the architect Robert Buckminster Fuller. Hence the name by which we now know this new class of molecules – fullerenes.

A letter to the scientific journal Nature in 1985, announcing their discovery, produced a sceptical reaction from many scientists. A peak in the mass spectrum was not considered enough evidence for a completely new form of carbon, largely because the amount to produce this peak is extremely small. Then in 1990 Wolfgang Kratschmer announced that he had prepared C_{60} by vaporising graphite in an inert atmosphere in a sufficient amount to allow him to analyse it. The structure was indeed the cage structure shown in Fig 52.

△ Fig 51 The mass spectrum of laser-vaporised graphite showing a large peak at 720 m/e.

△ Fig 52 Buckminsterfullerene, the first fullerene to be discovered.

QUESTIONS

1. Look at Fig 51.
   a) Why does the large peak suggest a species with the formula, C_{60}?
   b) The C_{60} species produced in the mass spectrometer is not a molecule. What is it and how is it formed?
   c) What species does the peak at 840 suggest?
   d) Why is the species at 720 considered to be stable?
2. a) Why did Kratschmer use an inert atmosphere to vaporise graphite rather than just air?
   b) Draw the structures of graphite and diamond. Use these to suggest a reason why Kratschmer did not use diamond.
3. In an experiment, 1.208 grams of graphite is vaporised to give pure C_{60} of 0.0306 milligrams.
   a) Express 0.0306 milligrams in grams using standard form.
   b) Calculate the percentage yield of C_{60}.
4. Explain why Kratschmer's experiment convinced scientists that C_{60} had indeed been produced by Kroto and his colleagues. [QWC]

Skills practised

1, 2, 3, 4, 5, 8, 9, 10, 12, 13, 17, 20, 25, 26, 27, 28, 29, 30, 31, 32

There are two Assessing Investigative Skills activities.

The questions help you to practise some skills in a practical context. They are similar to the practical assessments you will encounter on your course.

Assessing Investigative Skills

AIS1

One way to remove limescale from the heating element of an electric kettle is to use a dilute solution of sulfamic acid. One commercial product is a solution of this acid. It is tested by a student using volumetric analysis.

- The student uses a pipette and filler to transfer 25.0 cm³ of the sulfamic acid product into a 250 cm³ volumetric flask and adds distilled water up to the 250 cm³ mark.
- The same pipette is rinsed with the diluted solution of the sulfamic acid product.
- The pipette and filler are now used to transfer 25.0 cm³ of the diluted product into a conical flask.
- Thymolphthalein indicator is added to this solution and it remains colourless.
- A burette is filled with 0.100 mol dm⁻³ sodium hydroxide (NaOH(aq)).
- The endpoint is when one drop of sodium hydroxide solution turns the indicator pink.
- The burette readings are recorded to the nearest 0.05 cm³ in Table 11 below.

Titration number	1	2	3	4
Final burette reading/cm³	21.50	42.20	21.35	42.15
Initial burette reading/cm³	0.50	21.40	0.80	21.50

△ Table 11

Question 1

a) Draw this table and add a row to show the titres. (1)

b) Calculate the mean titre to the appropriate precision. (1)

The equation below represents the reaction that occurred. HA is sulfamic acid.

NaOH(aq) + HA(aq) → NaA(aq) + H₂O(l)

In the following calculations, show your working.

c) Calculate the amount, in moles, of NaOH used in the titration. (1)

d) Determine the amount, in moles, of sulfamic acid (HA) present in 25.0 cm³ of the diluted product solution. (1)

e) Calculate the amount, in moles, of sulfamic acid present in the 250 cm³ solution. (1)

f) Calculate the concentration, in mol dm⁻³, of sulfamic acid present in the original commercial product. Give your answer to three significant figures. (2)

g) What is:
   i) the maximum error in a single burette reading? (1)
   ii) the percentage error in one of the titre values in Table 11? (1)

h) When the pipette is used for the second time it is rinsed with the diluted solution the student had made.
   i) Explain why this is done. (1)
   ii) What would be the effect on the titre of not rinsing the pipette? (1)

i) Each time the student performs a titration she rinses out the conical flask with distilled water. What effect does this procedure have on the titre? (1)

j) An air bubble is present in the tip of the burette at the start of the first titration but not at the end. Explain why this makes the titre less reliable. (1)

Question 2

The student decides to determine the relative molecular mass of sulfamic acid. She uses the apparatus in Fig 70 to measure the volume of carbon dioxide produced when sulfamic acid reacts with 0.50 mol dm⁻³ sodium carbonate (Na₂CO₃) solution.

△ Fig 70 The apparatus used to measure the volume of carbon dioxide produced from the student's reaction.

The results she obtains are shown in Table 12.

Mass of conical flask and sulfamic acid	134.57 g
Mass of conical flask empty	132.87 g
Volume of sodium carbonate solution added	100 cm³
Volume of carbon dioxide collected	207 cm³

△ Table 12

a) What is the mass of sulfamic acid used? (1)

b) Using HA to represent sulfamic acid, write the balanced equation, including state symbols, for its reaction with sodium carbonate. (2)

c) How many moles of carbon dioxide are released? (1)

d) All the sulfamic acid reacts. How many moles of sulfamic acid are present in the weighed sample? (1)

Each question includes an indication of the number of points needed in your answer.

ANSWERS

The answers include helpful tips and hints.

Each QWC answer includes low, medium and high level answer pointers to help you improve.

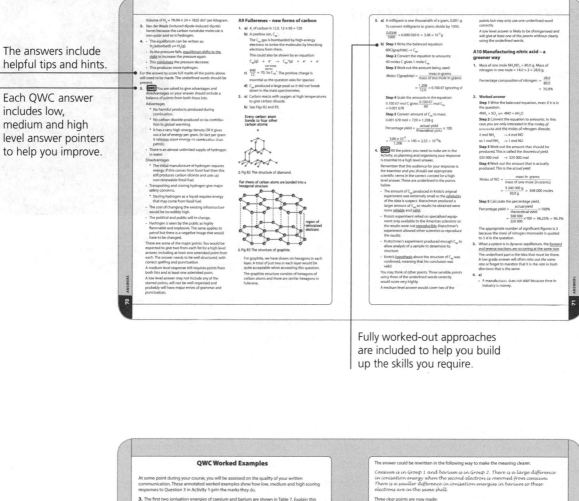

Fully worked-out approaches are included to help you build up the skills you require.

The QWC Worked Examples section shows low, medium and high written responses.

Comments explain how to improve the quality of written communication.

Skills

S1 EXPLAINING OBSERVATIONS

Science is a very reliable body of knowledge to which scientists continually add and, during your course, you will practise working scientifically.

Scientists:

- Observe. Observations are at the heart of how science develops and this is often called **data**.
- Question. From observations come questions, usually beginning with "Why?" or "How?" There might be a pattern in some data or something unexpected is recorded. If the same observations keep happening then the data is said to be **reliable**.
- Put forward ideas. The idea is often called a **hypothesis** and it seeks to explain what is causing something to occur as it does.
- Test hypotheses through experiments. **Predictions** can be made from a hypothesis and then experiments can be carried out to see if these predictions are correct. A very good example of testing a prediction is when Ernest Rutherford fired alpha particles at gold atoms (see A2).
- Devise theories. If the predictions are always correct, then the hypothesis is assumed to be correct and the explanation becomes accepted as a **theory**. A theory can include more than one hypothesis and it is used be used to make even more **precise predictions** that can be tested.
- Develop models. A **model** is another way of explaining observations. Sometimes it can be very simple, like thinking of an atom as containing thousands of negatively charged electrons embedded in a sphere of positive charge. This was even called the "plum pudding" model of the atom because the electrons were pictured as plums in a positively charged pudding (see A2). Sometimes the model can be very complex, like the way we now think of atomic orbitals or the π-bonds of benzene, both of which are based on mathematical models. A model often helps us to make sense of a theory. So long as the model or theory explains all known observations then it is accepted by the scientific community.

As you carry out experiments during your chemistry course you will produce results that are unexpected. There may well be a very good reason for this. You could have just picked up the wrong chemical! But when an unexpected result does not fit with the currently accepted theory, scientists have to decide what to do about it. The **anomalous result** (see S9) may be the result of experimental error or it could be the effect of some factor that is not properly understood that leads to a real change in our understanding.

A well-established theory is not usually rejected if it has proved to be useful in explaining particular phenomena. Sometimes models continue to be used because of their value in predicting particular behaviours. For example, it is still useful for us to think of electrons as particles in the outer shell around the nucleus of an atom because it accurately predicts the shapes of

△ Fig 1 Nobel Prize winner, Professor Sir Harold Kroto, holding buckminsterfullerene, C_{60}. See A9.

many molecules. You can see this in action in the electron pair repulsion theory in A6.

However, if anomalous results start to build up and seem to defy the current framework of scientific thinking then a different theory or model may explain observations better. Such a thing happened when it was realised that electrons could behave as waves as well as particles.

As you continue your study of chemistry you will see that familiar models and theories have sometimes been transformed by unexpected results. However, this is all part of the way scientists continually question, test and explain and it is this way of working scientifically that you will develop.

Apply this skill by doing Activities A1, A2, A3, A4, A5, A6, A8, A9, A11, A12, AIS2.

S2 UNDERSTANDING HOW SCIENCE ADVANCES

To achieve advancements in science, you need to draw from existing knowledge and understanding to pose new questions or suggest new scientific ideas. You might think of a new prediction to test an existing hypothesis and find that your prediction is incorrect, leading to a new hypothesis. This happened when the bond lengths in benzene molecules were accurately measured and did not fit the predictions of Kekulé's model. (See A11.) At other times new predictions may be correct and an existing model can be used in an entirely new way. For example, understanding the shapes of molecules allows the shapes of drug molecules to be predicted, leading to the design of new drugs which fit into receptor sites within the body. (See A6.)

△ Fig 2 This is a molecular model of a protein that helps to suppress tumour development in breast cancer.

Often science progresses faster when several scientists contribute to the development of new materials, ideas and theories. Having the appropriate understanding also means that scientists can challenge other scientists if they do not feel that the new ideas are correct.

Apply this skill by doing Activities A1, A2, A3, A5, A6, A9, A10, A11.

S3 USING APPROPRIATE TECHNIQUES

It is the observations that you make that lead to the hypotheses you put forward, if you are a scientist. This data is the evidence used to support ideas. *How* data is collected is extremely important, and this might mean devising a new procedure that produces reliable and precise data. If this is seen to be **valid** by the rest of the scientific community then it will be adopted as a protocol, which means an accepted procedure.

If you think of the impact that IT has on all our lives you would expect it to have an impact on science. In fact, it has

△ Fig 3 Selecting the correct techniques is essential when performing experiments.

revolutionised how scientists perform experiments and collect their data. Instrumentation has also improved so that data obtained is more reliable and more precise. For example, high-resolution mass spectrometry has allowed relative molecular masses to be calculated to four decimal places, so it is possible to determine which compound is present if accurate isotopic masses are used. You can try this yourself using Table 1.

Formulae	Accurate M_r (m/e values for M^+ ion)
C_4H_{10}	58.0780
C_3H_6O	58.0417
$C_2H_2O_2$	58.0054
$C_2H_6N_2$	58.0530

△ Table 1 Relative isotopic masses: ^{12}C = 12.0000; ^1H = 1.0078; ^{14}N = 14.0031; ^{16}O = 15.9949.

Linking up the techniques of gas-liquid chromatography and mass spectrometry provides a very effective method of analysing and identifying the different compounds in complex mixtures. Gas chromatography is an excellent way to separate components, whereas a mass spectrometer is not the instrument to use if there are too many different molecules in a mixture because the mass spectra are too complicated. However, if the mixture is first separated by gas chromatography then the mass spectra of the separated molecules are compared with a database of known mass spectra, a positive identification of a component can often be obtained. This combined technique is known as GC-MS (see A13).

Apply this skill by doing Activities A1, A6, A8, A9, A10, A11, A13, AIS1, AIS2.

S4 MAKING VALID OBSERVATIONS

Just because you make an observation it is not necessarily a valid one. When something does what it is meant to do it is **valid**. For the results of an investigation to be valid it must measure what it sets out to measure. This may seem obvious but sometimes there can be flaws in the way an experiment is designed or carried out, which means that the experiment is actually measuring something else entirely.

When iron-60 was first discovered on Earth in 1998 scientists were not convinced that the experiment was actually measuring the presence of iron-60 atoms because the amount was so incredible tiny (see A1). Similarly, when some noble gases were found to react it was not until other scientists had reproduced Neil Bartlett's experiment (see A5) that they were convinced that his procedure was valid.

Apply this skill by doing Activities A1, A2, A5, A9, A11, AIS1, AIS2.

S5 TAKING MEASUREMENTS
Accurate measurements

Accuracy is how close a measurement is to its true or accepted value. How accurate a result is depends on the quality of the apparatus used and the skill of the experimenter. If a thermometer is used in an experiment and it measures a temperature of 25 °C, and then another thermometer is substituted that gives the measurement as 26 °C under exactly the same conditions, then the accuracy of the thermometers is in doubt. To find out which thermometer is accurate we need to know the true value of the temperature being measured.

Fig 4a explains this by looking at arrows hitting a target. The closer an arrow is to the centre of the target the more accurate the shot.

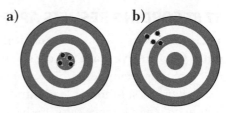

△ Fig 4 In a) the bow shots are accurate because they are in the centre circle of the target, and also precise because they are close together. In b) the bow shots are precise because they are close together but not accurate because they are not at the centre of the target.

Precise measurements

If measurements are close together then they are said to be precise. The closer the grouping of results the more precise they are. However, this does not mean they are accurate. So in Fig 4a the bow shots are precise because they are close together and, as we can see, they are also accurate. In Fig 4b the shots are still precise because they cluster together, but they are not accurate. The difference between these two terms, accurate and precise, can cause confusion and sometimes even examination questions appear to muddle their meanings.

Apply this skill by doing Activities A1, A2, A9, AIS1, AIS2.

S6 MANAGING RISK IN INVESTIGATIONS

When you carry out experimental work you should try to minimise the **risk** to yourself, others and the environment. To do this you must identify the hazards involved in the experiment you are about to perform and how likely each one is to occur. This is a process known as risk assessment. It should be conducted methodically and use available information.

A **hazard** is anything that is likely to cause harm and that includes harm to organisms in the outside environment. There are internationally recognised hazard warning symbols for most chemicals that are likely to be used, but it is not only chemicals that you should consider when managing risk.

- What are the hazards in the experimental procedure?
- If a flammable liquid is being heated, will the vapour come into contact with a naked flame?
- If a poisonous gas is being produced, is a fume cupboard being used?
- How are you going to dispose of your chemicals?

Apply this skill by doing Activities A2, A10, AIS1.

explosive   harmful to health   flammable

very toxic (poisonous)   harmful to breathing system   corrosive (attacks skin)

△ Fig 5 Some of the hazard pictograms that help scientists assess and manage risk.

S7 RECORDING RESULTS APPROPRIATELY

There are two types of data, **qualitative data** and **quantitative data**. **Qualitative data** are observations you can describe but not measure. For, example, in Fig 6a a yellow precipitate is produced in the colourless solution of lead nitrate and in Fig 6b a green flame is produced when copper salts are heated.

Quantitative data deals with quantities – observations you can measure. If you titrate a solution from a burette then the volume is measured in cm^3. If the percentage composition of a pollutant in air is measured, the number is followed by a % sign. The numbers always have to have something with them that tells you what they mean. Numbers with meanings are **values**.

How **data** is recorded and presented is very important. Recording data in a methodical way makes it easier to spot mistakes and identify **anomalous results**. It also makes it easier for other scientists to interpret your results. When writing down your results think about their margin of error. This is how much certainty you have in the measurement you have taken. A 50.00 cm^3 burette can be read to an accuracy of 0.05 cm^3 so you should quote your results to the nearest 0.05 cm^3. Thermometers can often be read to the nearest 0.1 °C.

There is always some uncertainty in measurement. When we talk about the accuracy of measurement we do not necessarily know the true value of that measurement but we should know how certain we can be of our measurements.

Tables are an excellent way to organise and present your information. If the table involves timing something then time is almost always written in the first column because it is the **independent variable**. Suppose you are measuring the volume of gas produced every 10 seconds in a reaction, to work out the rate. Your table would look similar to Table 2. Each column is labelled with what is being measured and the unit. The **dependent variable** is the volume of CO_2. It is called a dependent variable because the volume depends on how long the reaction has been going, Time is independent because it does not depend on the volume of CO_2 produced.

Notice that in Table 2, time is measured to the nearest second. If you were to put this value to one decimal place (0.0, 10.0, etc.) then you are claiming your result to a certainty of 0.1 s and this is probably incorrect. Likewise, when measuring the volume of gas with a measuring cylinder you can only be certain of a reading to the nearest whole number.

△ Fig 6a Potassium iodide solution is added to a solution of lead nitrate forming a yellow precipitate of lead iodide.

△ Fig 6b Copper in copper salts gives a green flame when heated in a Bunsen burner flame.

Time/s	Volume of CO_2/cm^3
0	0
10	40
20	80
30	115
40	125

△ Table 2 Volume of CO_2 produced in the reaction of calcium carbonate with 2.0 mol dm^{-3} HCl(aq).

Trial	1	2
Final burette reading/cm³ 26.4	27.45	27.25
Initial burette reading/cm³ 1.2	2.05	1.80
Titre/cm³ 25.8	25.40	25.45

Δ Table 3 If you are titrating using a burette then your results may look like this.

In this case the trial reading is a titration carried out quickly to see roughly where the **end point** is likely to be. The other burette readings are precise readings where the solution from the burette is added drop by drop until the end point is reached. The burette can be read to the nearest 0.05 cm³. In this case, because two precise values are within 0.1 cm³, we do not require any further readings (see S5).

Using graphs is another way to present data that can be extremely helpful when it comes to spotting any trends and patterns. We will look more closely at line graphs in S36.

Apply this skill by doing A3, A6, AIS1, AIS2.

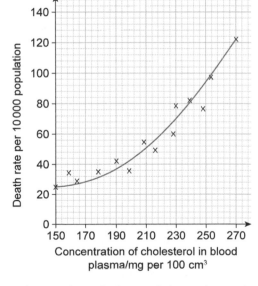

Δ Fig 7 Taking a burette reading. Notice that the eye is level with the meniscus. The reading is taken from the bottom of the meniscus.

S8 ANALYSING AND INTERPRETING DATA

Your experiment may have been designed to test a particular **hypothesis**. The data you obtain must be analysed and interpreted to see if it provides the evidence to support your hypothesis. When analysing quantitative data scientists are looking for any trends and patterns.

Interpreting data is what you need to do once you have analysed the results. Do the results support the hypothesis put forward? Could there be another explanation? Sometimes data will show a **correlation**. A correlation is when one factor appears to have a relationship with another. So, for example, there is a link between concentration of the cholesterol in blood and the death rate due to heart disease (see A14).

A **causal link** is when one factor directly affects another. Just because there is a correlation it does not mean there is a causal link. There may be another factor at work affecting both variables that is not immediately obvious. It could be that lack of exercise increases cholesterol levels in the blood and this then increases the percentage of deaths.

Scientists are more likely to accept a hypothesis that there is a causal link if a plausible mechanism exists to explain the link. The causal link in the depletion of the ozone layer was the production of CFCs (chlorofluorocarbons). The mechanism was

Δ Fig 8 Death rate in the population against total concentration of cholesterol in blood plasma.

the release of chlorine atoms from these molecules that went on to catalyse the reaction of ozone to produce oxygen.

Apply this skill by doing Activities A1, A2, A3, A4, A5, A9, A11, A12, A13, A14, A16, AIS1, AIS2.

S9 IDENTIFYING ANOMALOUS DATA

Data that does not fit an expected pattern, or lies outside the range of other results, is called **anomalous data**. How anomalous data is dealt with depends on whether there is an obvious cause, such as an error in your procedure. If there is then anomalous data can be ignored when analysing your results. However, if there is no obvious error a scientist will attempt to repeat the results. If the anomalous data is repeatable then it requires an explanation. When Rutherford fired alpha particles at gold atoms the anomalous result was the 1 in 10 000 particles that were reflected backwards and when this result was repeated he sought to explain it through his nuclear model of the atom (see A2). Other scientists may also seek to reproduce anomalous results. If they do, this will increase the confidence of the scientific community that something requires an explanation.

Apply this skill by doing Activities A1, A2, A4, A9, A11, A12, AIS1, AIS2.

S10 EVALUATING METHODOLOGY, DATA AND EVIDENCE

An evaluation of any science investigation is an assessment of how **reliable** the data is. When evaluating the **methodology** we need to consider the equipment, procedures, and techniques that have been used to produce the data. You should ask yourself when evaluating your own experiments: "Does the methodology used measure what I set out to measure?" This tells you if it is **valid**.

Data is considered reliable if it can be **repeated** and **reproduced**. This means that the same experimenter, using the same equipment, gets similar results. If the data is **reproducible** this means that other scientists can produce similar results using the same methodology.

△ Fig 9 A researcher analyses his data.

In evaluating data these are some of the questions a scientist asks:

- Is the methodology valid?
- Is the data **precise**? This means that the results are clustered close together.
- Are there any **anomalous** results? If so, is there a plausible explanation?
- Is the data **accurate** if true values are known?
- Are the results **reliable**?

Apply this skill by doing Activities A1, A2, A5, A9, A11, A12, A14, AIS1, AIS2.

S11 ASSESSING ERROR

When an experiment is repeated there will always be some slight variations in the data collected, however good you are as an experimenter. The **resolution** or **sensitivity** of the apparatus could well be a factor. These terms mean how well the apparatus detects small differences. The high-resolution mass spectrometer discussed in S3 can accurately measure relative atomic and molecular masses to four decimal places. The resolution of a 50.00 cm^3 burette is much higher than the resolution of a 50 cm^3 measuring cylinder.

Calculating percentage errors based on resolution of apparatus

There is always uncertainty in measurement: the higher the resolution, the less the uncertainty. This means the results will cluster around a smaller range of possible results.

Usually the smallest division on a scale is divided by two to give a value for its precision. So a 50.00 cm^3 burette has a graduated scale with the smallest division being 0.1 cm^3 so its precision is assessed as ± 0.05 cm^3. A balance weighing to two decimal places has a precision of $0.01/2$, so its error or uncertainty is ± 0.005 g.

From Table 3 in Skill 7, two burette readings, initial and final, are taken to calculate the titre. Each has an error of ± 0.05 so the error for this reading is $2 \times 0.05 = \pm 0.1$cm^3.

So in your calculation of percentage error for this titre:

$$\frac{2 \times 0.05}{25.45} \times 100 = 0.39\%$$

In the case of a balance reading to an accuracy of ± 0.005 g then one reading is taken. So if 5.04 g is the mass reading then the percentage error is

$$\frac{0.005}{5.04} \times 100 = 0.099\%$$

Researchers identify three types of error:

- **Human errors** are usually made by an inexperienced experimenter and can be as simple as using the wrong chemical or not waiting until all of a substance has reacted. In an examination you will not receive marks for just writing *human error* if you are asked to comment on why one procedure produces less accurate results than another. You must explain exactly what you think caused the particular error.
- **Random errors** are usually recognised when results are not precise and there is no pattern to them. There may be a fault in the technique or a limit to the resolution of the apparatus.
- **Systematic errors** are errors that produce inaccurate values. They are always in one direction, so results may be consistently high or consistently low. There can be many causes, such as a fault in the procedure that always affects the result in one particular way, or a piece of equipment that is measuring values that are too high or too low.

Apply this skill by doing Activities A1, A2, AIS1, AIS2.

S12 DRAWING VALID CONCLUSIONS

Having analysed the results, you can draw **conclusions**. Conclusions are your judgments. They are the decisions you make about how to interpret the data and they should be **valid**. Valid conclusions come from the data you have produced and only from the data you have produced. The data is your evidence: does it support the hypothesis and prediction you made when you planned your experiment? Are there anomalous results and can they be explained? If there are not, did you repeat the experiment to get similar results?

Apply this skill by doing Activities A1, A2, A5, A9, A11, A12, A14, A16, AIS2.

S13 COMMUNICATING SCIENTIFIC INFORMATION

When you write up an experiment you are communicating scientific information. This information needs to be understood by other scientists so the way you express yourself is important. Use this flowchart to check:

△ Fig 10

You are only doing what all scientists must do if they are going to share their research findings with the rest of the scientific community. The reasons scientists communicate their discoveries is so that other scientists can check that they agree with the findings made and the explanations given. This is why science is such a reliable body of knowledge and understanding. There are two main ways a scientist communicates with the rest of the scientific community:

1. through writing papers about their research for respected scientific journals

2. by reporting their findings at scientific conferences.

In both cases there is a process known as **peer review**. Other scientists (the peers) who are experts in a similar field of research are asked to look through the article or the conference proposal and do all the checks that we have already discussed. They also ask themselves questions about **bias**. (This is when results are shifted in a particular direction to support a hypothesis.)

- Are the results skewed because of poor experimental techniques so that they are not valid?
- Who paid the scientist to do the research? If it is a drugs company with a vested interest in getting approval for a new drug, or a food company wanting to refute a hypothesis about health risks of particular ingredients, then this could mean the scientist feels an obligation to report only results that will please an employer.
- What are the financial and other personal gains that the scientist stands to make?

It is the editor, or the conference organiser, who decides what happens next and they must ask similar questions on bias about the peer reviewers. But in the end it is the scientific community that validates the findings of other scientists by reproducing their results and putting the methodology under scrutiny.

Once published in a scientific journal, a particular piece of research may be relayed to the public via the media. When articles appear in newspapers, on the internet and on television and radio, the same care that scientists take when they report their findings and have them peer reviewed should also be taken by reporters, but this does not always happen. You are now in a position to judge these articles so ask yourself these questions:

- Is the evidence in the results valid?
- Does the evidence support the particular explanation being given?
- Is the terminology appropriate or is it being used in a confusing or confused way?
- Does the research show bias?

Apply this skill by doing Activities A1, A5, A9, A10, A14.

S14 CONSIDERING ETHICAL ISSUES

Ethics is judging the rightness and wrongness of our actions. It is a whole branch of knowledge on its own and there are different ways of approaching ethical questions. Ethics provides a framework to guide how we make decisions on a personal level, on a national level and internationally. It is an essential part of how we use our knowledge and understanding of science.

On a personal level it may be considering how a particular chemical is disposed of after an experiment by thinking of the potential consequences to the environment. On a global level it is considering whether we should use food for fuel when millions are starving, or whether food manufacturers should still use trans-fats in processed food (see A7 and A14). In fact most of the activities in this book raise ethical issues.

Apply this skill by doing Activities A7, A8, A10, A14.

S15 SCIENCE AND DECISION-MAKING

There are huge benefits for society from scientific advances and new technologies but there are also drawbacks and risks. As research continues, new risks are discovered meaning that some decisions need to be reversed. The way society makes decisions about science is through its decision-makers, often politicians, who may be influenced by:

- the financial cost
- the scientific evidence, which may be tentative, incomplete or even biased
- special interest groups
- public opinion, formed partly by how people understand the science involved
- their own vested interest
- the media
- ethics.

We can see this with the arguments about climate change and global warming. Most scientists do believe there is overwhelming evidence that average global temperatures are increasing but other scientists oppose this view. This makes the job of decision-makers across the globe more difficult and explains why international treaties are so hard to agree on.

However, scientists produced valid conclusions about the risks from carbon monoxide and nitrogen oxides released by car exhausts and the result was the introduction of catalytic converters, which dramatically reduced the potential harm to the environment, particularly in urban areas.

Apply this skill by doing Activities A7, A8, A10, A14.

S16 WRITING BALANCED EQUATIONS

Chemical equations use formulae. A **molecular formula** tells you exactly how many atoms of each element there are in a molecule. O_2 is an oxygen molecule made up of two oxygen atoms. But not all substances are made up of molecules. For example, sodium chloride is a giant lattice of sodium ions and chloride ions. The number of sodium ions is the same as the number of chloride ions and so we use the formula $NaCl$ to indicate this 1:1 **ratio**. For these compounds it is more correct just to use the word term **formula** or **formula unit**.

We can learn more about balancing equations by looking at this reaction.

When carbon burns in oxygen it can form carbon monoxide if there is insufficient oxygen present to form carbon dioxide. The equation is:

$$C + O_2 \rightarrow CO$$

reactants   product

This equation is unbalanced. There are more oxygen atoms on the left side of the equation than on the right. An atom of oxygen has disappeared and this is not possible in chemical reactions.

You might be tempted to put CO_2 on the right-hand side to balance the equation but now you have changed the carbon monoxide to carbon dioxide, a different gas with different properties. Instead, as shown in Fig 11, we balance the equation for the reaction by using two carbon atoms to produce two molecules of CO.

$$2C + O_2 \rightarrow 2CO$$

Chemical equations often include **state symbols**, which tell you the physical state of the reactants and products at room temperature, unless another temperature is stated.

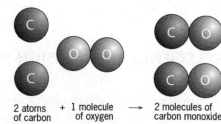

2 atoms   + 1 molecule   →   2 molecules of
of carbon   of oxygen   carbon monoxide

△ Fig 11 Combining two carbon atoms with one oxygen molecule.

(s) – solid (l) – liquid (g) – gas (aq) – in **aqueous solution**

So the balanced equation may be written:

$$2C(s) + O_2(g) \rightarrow 2CO(g)$$

Apply this skill by doing Activities A4, A8, A12, A16, AIS1.

S17 RELATIVE ATOMIC MASS

The masses of atoms are compared using the relative atomic mass scale or A_r scale. On this scale the relative atomic mass of carbon-12 is the standard and the A_r values of all other atoms are measured relative to this standard.

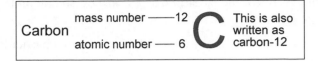

Carbon   mass number ——12   **C**   This is also written as carbon-12
atomic number —— 6

△ Fig 12 Carbon-12 showing the mass number and atomic number.

Relative atomic mass of an element is the average mass of an atom (taking into account all its isotopes and their abundance) compared to 1/12 the mass of one atom of carbon-12.

Isotopes are atoms of the same element, which means they have the same atomic number (proton number) but different **mass numbers**.

Mass number is the total number of protons and neutrons in an atom.

Mass spectrometers are instruments used to calculate relative atomic masses very accurately. By using the mass spectrum of neon we can see how to determine the relative atomic mass of neon.

There are three naturally occurring isotopes of neon, ^{20}Ne, ^{21}Ne and ^{22}Ne and these show up as peaks on the mass spectrum (see Fig 13). The y-axis tells us the percentage abundance of each of these isotopes. The x-axis requires more explanation. The mass spectrum actually measures the mass of the positive ions it creates. For the purposes of your course we consider only the ions with a single positive charge: ^{20}Ne$^+$, ^{21}Ne$^+$ and ^{22}Ne$^+$. The charge is therefore +1, so the mass/charge ratio is the same as the mass because we are dividing by +1.

The percentage abundance tells you how many atoms of each isotope are present in every 100 atoms.

The A_r is calculated as follows:

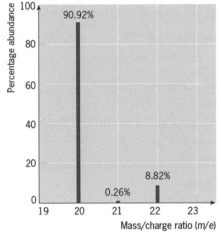

△ Fig 13 The mass spectrum of naturally occurring neon, showing the percentage abundance of each isotope.

$$A_r \text{ neon} = \frac{(90.92 \times 20) + (0.26 \times 21) + (8.82 \times 22)}{100} = \frac{2018}{100}$$

$$= 20.18$$

Apply this skill by doing Activities A4, A8, A12, A16, AIS1.

S18 AMOUNTS AND THE MOLE

When chemists use the word **amount**, they are talking about the number of particles in a substance. The particles can be atoms, molecules, ions, electrons, etc. To describe a number of eggs, you use the unit dozens. As a chemist, when you talk about amount you use the unit **moles** (symbol **mol**).

One mole is the amount of substance that contains as many particles as there are atoms in exactly 12 g of carbon-12.

The relative atomic mass of magnesium is 24. This means that one atom of magnesium has twice as much mass as one atom of carbon-12, with an A_r of 12.

There is one mole of carbon atoms in 12 g so one mole of magnesium atoms is 24 g.

To work out how many moles of atoms are in a particular mass of substance, use this equation:

$$\text{amount in moles} = \frac{\text{mass in grams}}{\text{mass of one mole (in grams)}}$$

Example

Q How many moles of atoms are there in 56 g silicon? (A_r of Si = 28)

A amount in moles $= \dfrac{\text{mass in grams}}{\text{mass of one mole (in grams)}}$

$\qquad$ moles of silicon $= \dfrac{56 \text{ g}}{28 \text{ g}}$

$\qquad\qquad\qquad\qquad = 2 \text{ mol}$

Q What is the mass of 0.25 mol calcium? (A_r of Ca = 40)

A mass in grams $\quad$ = amount in moles $\times$ mass of 1 mole

$\quad$ mass of calcium = 0.25 mol $\times$ 40 g

$\qquad\qquad\qquad\quad$ = 10 g

One mole of a substance is 6.02×10^{23} particles. This is a gigantic number (see Skill 29 about **standard form**). There is not even one mole of sand grains around the whole of the coastline of Africa.

The number of atoms or molecules in one mole is also known as the Avogadro constant.

Apply this skill by doing Activities A4, A8, A10, A11, A12, A14, AIS1, AIS2.

S19 RELATIVE MOLECULAR MASSES

Carbon dioxide is made up of molecules each having the **molecular formula** CO_2.

Fe_2O_3 is not made of molecules but this is its **formula unit** or **formula**.

The **relative molecular mass M_r** is calculated using the relative atomic mass scale. Again we use carbon-12 as the standard against which the masses of molecules or formula units are compared.

The M_r of CO_2 $\quad = A_r$ of C $+ (A_r$ of O $\times 2)$

$\qquad\qquad\qquad = 12 + (16 \times 2) \qquad = 44$

The M_r of Fe_2O_3 $\quad = (A_r$ of Fe $\times 2) + (A_r$ of O $\times 3)$

$\qquad\qquad\qquad = (56 \times 2) + (16 \times 3) \quad = 160$

△ Fig 14 Working out the relative molecular mass of carbon dioxide.

In this case M_r refers to the formula Fe_2O_3 and is called the **relative formula mass**.

During your course you will come across formulae with brackets, for example $Ca(OH)_2$. Fig 15 shows what this means.

Notice that the particles in $Ca(OH)_2$ are ions and have positive and negative charges. The charges on ions do not affect their A_r values.

So:

M_r of $Ca(OH)_2 = A_r$ of $Ca + [2 \times (A_r$ of $O + A_r$ of $H)]$

$= 40 + [2 \times (16 + 1)]$

$= 40 + (2 \times 17) = 74$

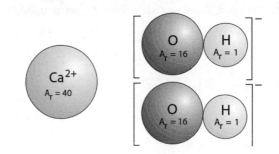

△ Fig 15 Diagram to explain the parts of the formula $Ca(OH)_2$.

Now that we can work out the relative masses of compounds we can use this to calculate masses in reactions using S20 Calculating Amounts and Masses from Chemical Equations.

Apply this skill by doing Activities A1, A7, A8, A10, A12, A13, A14, AIS1, AIS2.

S20 CALCULATING AMOUNTS AND MASSES FROM CHEMICAL EQUATIONS

Chemical equations are extremely useful because they allow us to predict the masses we can expect to obtain from reactions (**theoretical yield**) and how much reactant we need to use.

Let's look at the reaction with carbon and oxygen to produce carbon monoxide.

$2C$ $+$ O_2 $\rightarrow$ $2CO$

This means:

2 atoms C $+$ 1 molecule O_2 $\rightarrow$ 2 molecules CO

and it also means:

2 mol C atoms $+$ 1 mol O_2 molecules $\rightarrow$ 2 mol CO molecules

Carbon monoxide produced in a blast furnace reduces (removes oxygen from) the iron(III) oxide to iron.

$Fe_2O_3(s) + 3CO(g) \rightarrow 2Fe(s) + 3CO_2(g)$

Suppose we want know how much iron can be produced from 10 tonnes of Fe_2O_3. When doing calculations like this it is always best to show clearly the steps in your calculation.

Step 1 Write the balanced equation.

Do this even if the equation is given in the question because it will help you to organise your answer.

$Fe_2O_3(s) + 3CO(g) \rightarrow 2Fe(s) + 3CO_2(g)$

Step 2 Convert the equation to amounts.

In this example we do not need to know the moles of CO or CO_2.

1 mol Fe_2O_3 gives 2 mol Fe

Step 3 Work out the amount being used.

Mass of 1 mol $Fe_2O_3 = (56 \times 2) + (16 \times 3) = 160$ g

Remember 1 tonne is 1 000 000 g.

Moles of Fe_2O_3 in 10 tonnes moles $= \dfrac{\text{mass in grams}}{\text{mass of one mole (in grams)}}$

$$= \dfrac{10\ 000\ 000\ \text{g}}{160\ \text{g}}$$

$$= 62\ 500\ \text{mol}$$

Step 4 Scale the amounts in the equation.

1 mol Fe_2O_3 produces 2 mol Fe

62 500 mol Fe_2O_3 produces 125 000 mol Fe

Step 5 Convert amount (moles) of Fe to a mass.

125 000 mol × 56 g = 7 000 000 g

= 7.0 tonnes of Fe (to two significant figures)

So the theoretical yield for this reaction is 7.0 tonnes.

As you go through your course, you will find that you will develop this skill because you will use it many times.

Apply this skill by doing Activities A8, A9, A10, A12, AIS1, AIS2.

S21 MOLAR GAS VOLUMES

In 1811, the Italian scientist Avogadro put forward a hypothesis that:

Equal volumes of different gases contain equal numbers of molecules at the same temperature and pressure.

It is called Avogadro's law because all gases at low pressure obey it.

This means that the volume of one mole of any gas – the **molar gas volume** – must be the same under identical conditions of temperature and pressure.

At 298 K and 101 kPa (sometimes called room temperature and pressure) the molar gas volume is 24 dm³.

Remember that 298 K is 25 °C because 0 °C is 273 K (273 + 25 = 298 K).

kPa is kilopascals and a kilopascal is 1 000 Pa.

dm³ stands for decimetre cubed. A decimetre is 10 cm, so a decimetre cubed is 10 cm ×10 cm ×10 cm = 1000 cm³.

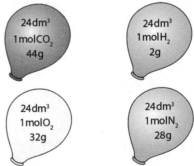

△ Fig 16 The molar volume of four gases at 298 K and 101 kPa.

Example

Q What is the volume at room temperature and pressure (rtp) of:

a) 1 mol hydrogen; **b)** 2 mol hydrogen chloride; **c)** 0.5 mol oxygen?

A a) Since 1 mole of any gas occupies 24 dm³ at rtp, the volume occupied by 1 mole of hydrogen is 24 dm³.

b) The amount of gas is 2 moles, so the volume is double the molar volume at rtp: $24 \times 2 = 48$ dm³.

c) Oxygen will occupy half the molar volume since there is 0.5 mol present: $24 \times 0.5 = 12$ dm³.

Understanding how to work out volumes from moles allows us to use equations to predict the volumes of gases being released from reactions.

Apply this skill by doing Activities A8, A12, AIS2.

S22 CALCULATING VOLUMES OF GASES FROM CHEMICAL EQUATIONS

Just as in S20, you can see just how useful chemical equations are. This time we use equations to predict the volumes of gas produced in reactions. Again, approaching these calculations methodically and setting down the steps is the key to exam success.

Let's try an example.

Example

Q What volume of carbon dioxide is produced when 2.90 kg of butane is burnt under room conditions (rtp)?

A Step 1 Write the balanced equation for the reaction:

$C_4H_{10}(g) + 9O_2(g) \rightarrow 4CO_2(g) + 10H_2O(l)$

Step 2 Convert the equation to amounts:

1 mol C_4H_{10} gives 4 mol CO_2

Note: In this example we do not need to know the number of moles of O_2 or H_2O.

Step 3 Work out the amount being used:

Mass of 1 mol $C_4H_{10} = (12 \times 4) + (1 \times 10) = 58$ g

(Remember 1 kg is 1000 g.)

$$\text{Moles of } C_4H_{10} \text{ in 2900 g} = \frac{\text{mass in grams}}{\text{mass of one mole (in grams)}} = \frac{2900 \text{ g}}{58 \text{ g}}$$
$$= 50.0 \text{ mol}$$

Step 4 Scale the amounts in the equation:

1 mol C_4H_{10} produces 4 mol CO_2

50.0 mol C_4H_{10} produces 200 mol CO_2

Step 5 Calculate the volume of gas:

1 mole of gas occupies 24 dm³ under room conditions

200 mol $\times$ 24 dm³ = 4800 dm³

So the volume of carbon dioxide produced = 4800 dm³.

Apply this skill by doing Activities A8, A12, AIS2.

S23 WORKING WITH AMOUNTS IN AQUEOUS SOLUTION

Water covers most of our planet and many reactions take place in aqueous solution. Nearly all the chemical reactions in our bodies take place in the aqueous environment of our cells. Being able to work out amounts in solution is very important and you will find that your course will expect you to understand how to calculate amounts in solution.

We need to understand a few terms that you have may have already met.

- **Solute** is the substance that is dissolved.
- **Solvent** is the liquid that does the dissolving and in your course this will usually be water.
- **Solution** is what is produced when the solute dissolves in the solvent.
- **Aqueous solution** is a solution where the solvent is water and in equations the state symbol is (aq).
- **Concentration** can be expressed as the mass of substance in a certain volume. The units chemists often use are *grams per cubic decimetre or g dm³*.

40 grams of sodium hydroxide in one cubic decimetre of aqueous sodium hydroxide is 40 g dm⁻³.

Remember a cubic decimetre is 1000 cm³.

Chemists, however, are usually interested in the *amount* of substance present in a solution. This is because once you know how many moles of solute are present in a particular volume of solution then you know how many particles are there.

The concentration of a solute in a solution is measured in *moles per cubic decimetre – mol dm⁻³*.

Examples

Q Work out how many moles of solute are dissolved in these solutions:

a) 25.0 cm³ of 1.0 M NaOH(aq)

b) 20.4 cm³ of 0.50 mol dm⁻³ HCl(aq)

A a) Number of moles NaOH in 1000 cm³ (1 dm⁻³) = 1 mol

$$\text{So number of moles NaOH in 25.0 cm}^3 = \frac{1 \times 25.0}{1000}$$
$$= 0.0250 \text{ mol}$$

b) Number of moles HCl in 1000 cm³ = 0.50 mol

$$\text{So number of moles HCl in 20.4 cm}^3 = \frac{0.50 \times 20.4}{1000}$$
$$= 0.0102 \text{ mol}$$

Q What is the concentration (in mol dm⁻³) of the following aqueous solutions?

a) 0.25 mol H_2SO_4(aq) dissolved in 50 cm³

b) 5.85 g NaCl dissolved in 250 cm³

A a) There are 0.25 mol in 50 cm³.

$$\text{So in 1000 cm}^3\text{, the number of moles} = 0.25 \times \frac{1000}{50}$$
$$= 5.0 \text{ mol}$$

Concentration of H_2SO_4 in solution = 5.0 mol dm⁻³

b)

$$\text{Moles of NaCl in 5.85 g} = \frac{\text{mass in grams}}{\text{mass of one mole (in grams)}} = \frac{5.85 \text{ g}}{58.5 \text{ g}} = 0.100 \text{ mol}$$

There are 0.100 mol NaCl in 250 cm³

$$\text{So in 1000 cm}^3\text{ the number of moles} = 0.100 \times \frac{1000}{250} = 0.400 \text{ mol}$$

Concentration of NaCl solution = 0.400 mol dm⁻³

Apply this skill by doing Activities AIS1, AIS2.

S24 TITRATIONS

A **titration** is an experimental technique used by chemists to find out the amounts of substances that are reacting in known volumes of solution very accurately. The technique is often referred to as **volumetric analysis**.

In a titration:

- one of the solutions is of known concentration
- the balanced equation tells us how many moles of each substance will react
- we can calculate the concentration of the other solution.

In an acid-base titration:

- acids neutralise bases
- the reaction is followed using an **indicator**
- the indicator changes colour at the **end point** or **neutralisation point**
- the end point is when the acid exactly neutralises the base.

See Table 3 in S7 for an example of how to record titrations.

△ Fig 17 The base solution (alkali) has been coloured using phenolphthalein indicator. It is almost at the end point, which will be when the addition of one drop of acid produces a colourless, neutral solution.

Example

Q What is the concentration of a dilute solution of hydrochloric acid if 26.1 cm³ is exactly neutralised by 25 cm³ of 0.100 mol dm⁻³ sodium hydroxide?

A Step 1 Write the balanced equation:

$NaOH(aq) + HCl(aq) \rightarrow NaCl(aq) + H_2O(l)$

Step 2 Convert the equation to amounts:

1 mol NaOH + 1 mol HCl → 1 mol NaCl + 1 mol H_2O

Step 3 Work out the amount of sodium hydroxide being used:

Number of moles NaOH in 1000 cm³ = 0.100 mol

So number of moles NaOH in 25.0 cm³ = $0.100 \times \dfrac{25.0}{1000} = 0.0025$ mol

Step 4 Scale the amounts in the equation:

1 mol NaoH reacts with 1 mol HCl

0.0025 mol NaOH reacts with 0.0025 mol HCl

Step 5 Scale the amount in the known volume to 1 dm³:

0.0025 mol HCl in 26.1 cm³

In 1 dm³ (1000 cm³) moles HCl = $0.0025 \times \dfrac{1000}{26.1}$

Therefore concentration of HCl = 0.0958 mol dm⁻³ (to 3 significant figures)

You will have the chance to look at another titration when you apply your skill in Activity AIS1.

S25 WRITING FOR YOUR INTENDED AUDIENCE

This is a very important skill, whether you are sitting an examination or applying for a job. Ask yourself these questions:

△ Fig 18 Often your intended audience is the examiner.

- Why am I writing this? You might have been asked to explain a chemical concept or justify a particular reaction mechanism.
- Who is going to read it? This is your audience and it will often be the examiner. However, you might be asked to write a newspaper article or put forward arguments to a public enquiry. Think about using appropriate chemical terms and how you will explain these.
- What form should it take? Sometimes this will include diagrams, tables or graphs.
- What information should I select? Think about what the question is asking and make sure that all the points you make are relevant.
- Is my writing style appropriate? You should write in a formal way, without the use of slang.

Apply this skill by doing Activities A1–A16.

S26 ENSURING MEANING IS CLEAR

To ensure your answers are easy to understand, ask yourself these questions:

- Are other people, such as examiners, going to able to read my writing?
- Have I spelled chemical terms correctly? Sometimes by misspelling these you can muddle up their meaning. You might write *neutrophile* (not a chemical term) when you really mean *nucleophile*. This would make an examiner wonder if you really think that neutrons are involved in a reaction mechanism.

- When writing in full sentences, is my grammar correct? Avoid making basic errors like missing out full stops and not starting sentences with capital letters.

For example, this sentence is confusing:

The second ionisation energy for magnesium is lower than for sodium the second electron is removed from the same shell and this requires less energy in magnesium.

The writer has not made it clear whether the second electron being removed comes from sodium or magnesium. The insertion of a capital letter and full stop makes this clear and should gain a mark.

The second ionisation energy for magnesium is lower than for sodium. The second electron is removed from the same shell and this requires less energy in magnesium.

Apply this skill by doing Activities A1–A16.

S27 ORGANISING INFORMATION CLEARLY AND COHERENTLY

This is a skill that is worth developing from the very beginning of your course because it will help you organise exam answers when you are under time pressure. The best place to start practising is with the notes you take in lessons; this will also make them easier to understand during revision. Check through them and try to organise the information before you forget what they were supposed to mean.

When it comes to answering exam questions that require longer answers, try to do the following:

- Read the question carefully and check that you understand what you are being asked to do. Identifying the key words will help you.
- Make a rough plan by jotting down the points you need to make. Do not spend too long on this.
- Decide how you will present the information. It may be that a table, graph or diagram is the clearest way to get some of the information across.
- Diagrams need to be clearly labelled. Reaction mechanisms requiring curly arrows are a good example. These need to be carefully drawn so that you show your understanding of them.

A useful tip when revising is to use past paper questions and make plans of how you would answer them, then check your plan with the published mark scheme.

Apply this skill by doing Activities A1–A16.

S28 USING SPECIALIST VOCABULARY

You will be expected to show in your answers that you can use chemical terms in the correct context to show that you do understand them. This is not just a case of defining them, although you should ensure that you are able to do this.

Look at this sentence:

Benzene undergoes *electrophilic substitution* reactions whereas alkenes react with *electrophiles* through *addition* reactions.

Depending on how far through your course you are, you may not understand all the words in italics, but by the end of your course you will need to be able to construct answers that show you understand the meaning of these terms. When you revise make sure you understand the specialist vocabulary you have covered. Making a glossary is a good way to do this. Also try to make up new sentences that include the specialist terms you are revising.

Apply this skill by doing Activities A1–A16.

S29 STANDARD FORM

This is a really useful way of writing very large and very small numbers.

Standard form can be shown as

$a \times 10^n$ (see Fig 19)

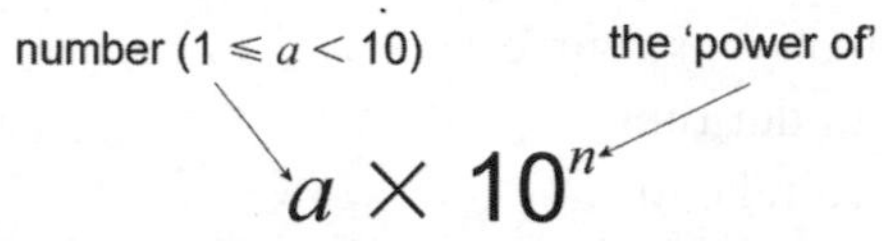

number $(1 \leqslant a < 10)$ the 'power of'

$$a \times 10^{n}$$

△ Fig 19

a is a number equal to, or bigger than, 1 and less that 10.

n tells us how many places to move the decimal point. It is also known as *the power of* and in this case it is the power of 10. n is also called the exponential.

There are 1 000 000 g in a metric tonne. To write this in standard form, take the first number on the left, which is 1. (This is a.) We now have to move the decimal point six times to reach 1, so n is 6, as in Fig 20.

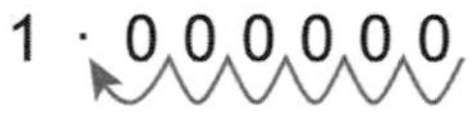

$1 \cdot 0\,0\,0\,0\,0\,0$

△ Fig 20

So in standard form we write 1×10^6.

Suppose we have 5.42 tonnes of magnesium produced by a manufacturer. In grams this is 5 420 000. The figure on the left is 5 and we have to move the decimal point six times to reach this number.

So in standard form we write:

5.42×10^6 g

This tells us that we multiply 5.42 by $10 \times 10 \times 10 \times 10 \times 10 \times 10$.

The mass of a carbon-12 atom in grams is

0.000 000 000 000 000 000 000 019 9 g.

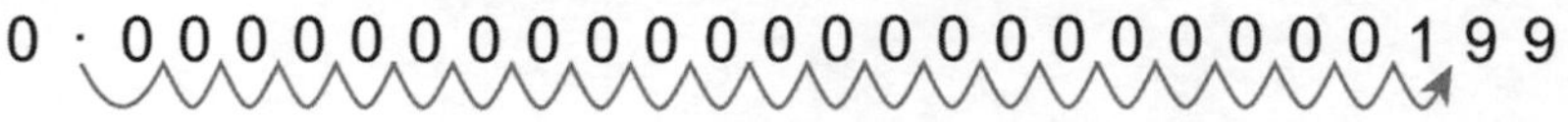

$0 \cdot 000000000000000000000000199$

△ Fig 21

So because we have to move the decimal point 23 times to the right, as in Fig 21, to get 1.99 in standard form, this is:

1.99×10^{-23} g

This is an incredibly small number and quite inconvenient to write. It is no wonder that we use relative masses of atoms rather than actual masses in grams – see S17.

If you wish to convert a decimal number to standard form then you reverse the procedure:

1.32×10^4 – you move the decimal point four places to the right, as in Fig 22:

$$1 \cdot 3\,2\,0\,0$$

△ Fig 22

$= 13\,200$

1.34×10^{-3} – this time the decimal point shifts three places to the left, as in Fig 23:

$$0\,0\,1 \cdot 3\,4$$

△ Fig 23

$= 0.001\,34$

Adding and subtracting numbers in standard form

For example:

5.65×10^3 g $+ 6.78 \times 10^5$ g

When adding or subtracting you have to be careful not to make a mistake if the powers of 10 are different.

Convert the numbers into the same power of 10.

0.0565 g $\times 10^5$ g $+ 6.78 \times 10^5$ g

Add or subtract the decimal number. In this case we add.

$(0.0565 + 6.78) \times 10^5$

$= 6.8365 \times 10^5$ g or 6.84×10^5 g to 3 significant figures (see S32).

Multiplying and dividing numbers in standard form

When multiplying numbers in standard form we add the powers of 10:

$10^4 \times 10^{-6} = 10^{4 + (-6)} = 10^{-2}$

When dividing, we subtract the powers:

$\dfrac{10^4}{10^{-6}} = 10^{4 - (-6)} = 10^{10}$

Examples

Q Calculate the total moles of 3.42×10^3 mol $\times 2.48 \times 10^6$ mol.

A Multiply $3.42 \times 2.48 = 8.4816$

Add the powers of $10^{(3+6)} = 10^9$

8.4816×10^9 mol $= 8.48 \times 10^9$ mol to 3 significant figures

Q Calculate the number of kilojoules

$$\frac{7.40 \times 10^4 \text{ kJ}}{3.64 \times 10^5 \text{ kJ}}$$

A Divide the decimal numbers: $\dfrac{7.40}{3.64} = 2.032\,967\,033$

Subtract the powers $10^{4-5} = 10^{-1}$

$2.032\,967\,033 \times 10^{-1}$ kJ $= 2.03 \times 10^{-1}$ kJ to significant figures

Apply this skill by doing Activities A1, A8, A9, A11, A12, A15, A16.

S30 RATIOS, FRACTIONS AND PERCENTAGES

A **ratio** is a way of comparing two quantities. Naturally occurring chlorine has two isotopes, ^{35}Cl and ^{37}Cl. These are in a ratio of 3:1. This means that there are three times as many ^{35}Cl atoms as there ^{37}Cl atoms.

A **fraction** is one number divided by another.

In the case of the abundance of chlorine isotopes, $\frac{3}{4}$ of the atoms are ^{35}Cl atoms.

- The total amount of naturally occurring chlorine atoms has been spit into four parts. This number goes on the bottom of the fraction and is called the denominator.
- Three of these four parts are due to ^{35}Cl and this is the number that goes on the top of the fraction and is called the numerator.

So a fraction tells you how much of something there is.

If there is 1 mole of NaOH in 1000 cm^3 then in 25 cm^3 there are 25/1000 of 1 mole.

A **percentage** is a fraction that has 100 on the bottom of it, so 100 is the denominator. So 3% is 3/100.

Example

Q Some Egyptian crude oils contain 3% sulfur. How much sulfur is in 15 tonnes of crude oil?

A 15 tonnes $\times \dfrac{3}{100} = 0.45$ tonne of sulfur

To show a fraction as a percentage, multiply the fraction by 100.

For example, $\frac{3}{4}$ of all naturally occurring chlorine atoms are ^{35}Cl atoms.

So the percentage of ^{35}Cl atoms is:

$$\frac{3}{4} \times 100 = 75\%$$

Apply this skill by doing Activities A1, A8, A9, A11, A12, A15, A16.

S31 USING YOUR CALCULATOR

Calculators can vary so some functions may look slightly different depending on how your keys are designed. Please check with your teacher if you do not understand what your keys mean.

Standard form calculations

In S29 we performed calculations using standard form. It is much more convenient to use a calculator.

To multiply 3.1×10^1 by 6.8×10^{-3} (see Fig 24).

Using logarithms

The most common logarithm you are likely to meet is log to the base 10 or $\log_{10}$ and this is often just written as log.

Suppose you want to convert ionisation energy to $\log_{10}$ of ionisation energy (see A2).

△ Fig 24

The tenth ionisation energy of sodium is 141 000 kJ mol^{-1}.

To convert this to $\log_{10}$

- Press key
- Enter 141 000
- Press ▤
- The display will read 5.149219113, which is 5.149 to 3 significant figures or sf (for significant figures of logarithms we count the numbers to the right of the decimal point).

What this really means is 10 to the power 5.149, which is $10^{5.149}$

Obtaining antilogarithms

When you want to convert $\log_{10}$ to a decimal number this is called obtaining the antilogarithm (also known as an *inverse logarithm*).

Let's convert $\log_{10}$ 5.149 back to a decimal number.

- Press the SHIFT or 2ndF key to give you access to the second functions on your calculator. These are the functions that are written above your calculator keys.
- Press the 10ˣ key
- Enter 5.149
- Press ▤
- The display will read 140928.8798, or 141 000 to 3 sf.

You use the same procedure if you want to convert a negative log to a normal decimal number.

Give the standard form value of $\log_{10} = -3.674$

- Press the SHIFT or 2ndF
- Press the 10ˣ
- Enter −3.674
- press =
- The display will read 0.000211836.
- 2.12×10^{-4} in standard form to 3 sf

$-\log_{10} 3.674$ means $10^{-3.674}$

At first sight you might expect an antilog of something $\times 10^{-3}$ but with negative logarithms the power of the antilog is to the next whole number, in this case $\times 10^{-4}$.

What your calculator is doing is this:

$10^{-3.647} = 10^{(1-0.674)} \times 10^{-4} = 10^{0.326} \times 10^{-4}$.

Squaring, cubing and other powers of numbers

To square a number enter the number and press the x^2 key followed by =. So $2^2 = 4$.

To cube a number use the x^3 key. $2^3 = 8$. Try this for yourself.

If you want to work out 2^4 you may have a y^x key or x^y key or ^.

Enter 2 then y^x (or whichever alternative key you have) then 4 and press =. So $2^4 = 16$.

Estimating

It is very easy to make a mistake keying in values on a calculator so it is always a good idea to have an estimate of the order of magnitude to expect.

Note: When we talk about *order of magnitude* we are usually referring to the powers of 10. For example, 10^1 to 10^2 is an increase of one order of magnitude.

Whatever you are multiplying or dividing, round off the values to 1 significant figure. This makes the calculation much easier and you will probably be able to do it in your head. For more about significant figures, see S32.

Estimating the answers using normal decimal numbers:

$\dfrac{2046}{0.482} = \dfrac{2000}{0.5} = 4000$ (The actual answer is 4240 to 3 sf)

Estimating calculations that multiply or divide numbers in standard form:

- $(6.02 \times 10^{-3}) \times (4.23 \times 10^{-3})$
- Round numbers up to 1 sf
- $6 \times 4 = 24$
- Add the powers of 10 together (or if you are dividing, subtract them):
  $10^{-3+(-3)} = 10^{-6}$
- $= 24 \times 10^{-6}$, which is 2.4×10^{-5} (actual answer to 3 sf $= 2.55 \times 10^{-5}$)

Even when you are really pushed for time in an examination, just adding or subtracting powers will give you some guide as to whether you have the correct order of magnitude for your calculation.

Apply this skill by doing Activities A1, A3, A4, A7, A8, A9, A10, A11, A12, A13, A14, A15, A16, AIS1, AIS2.

S32 USING AN APPROPRIATE NUMBER OF SIGNIFICANT FIGURES

Whenever you take a measurement there will be some uncertainty in that measurement. We first looked at this in Skill 11. If you are reporting your results or doing a calculation then you need to give other scientists an idea of how precisely you are able to take a particular measurement. This is where **significant figures** come in.

Significant figures are a scientist's way of telling you how precise the measurements are. Remember **precise** means how close together they are. The more uncertain a measurement the less precise results may be.

Determining the number of significant figures

Here are a few simple rules to help you determine the number of significant figures – often abbreviated to sf.

- Count the number of digits from left to right starting with the first digit that is not zero.

0.005670 has 4 sf because there are four digits including the first non-zero.

Notice that we count zero if it lies to the right of the first non-zero digit.

56.700 has 5 sf.

- When a number is in standard form ignore the exponential (that is, 10 to the power of):

1.35×10^{-6} is to 3 sf, whereas 9.670×10^6 is to 4 sf.

- When dealing with **logarithms** you count the numbers to the right of the decimal point.

Log_{10} 4.673 is to 3 sf, *not* 4. This is because it means $10^{4.673}$.

These rules do not apply to whole numbers. If you have a value of 100 m you cannot tell if this is 1, 2 or 3 sf.

Some numbers are exact. For example there are exactly 1000 cm^3 in 1 dm^3 so the number of significant figures is not relevant.

Decimal places

When looking at a number, count the digits including zeros to the right of the decimal place (dp). Sometimes an examination question asks you to quote a value to a certain number of decimal places. If a question asks you to do this, don't confuse it with the number of significant figures. Decimal places and significant figures are very often different.

Significant figures in calculations

This section will help you to be sure of the number of significant figures you should give in a calculation.

Rounding is the process by which you reduce the number of significant figures in your answer.

- Decide on the number of significant figures you need.
- Look at the number that comes immediately after the last significant figure on the right.
- If it is 4 or less then delete all the others.

So 0.83432 is 0.834 to 3 sf.

- If it is 5 or more then add 1 to the number that comes next to it.

So 12.756 is 12.8 to 3 sf.

Adding and subtracting: find the value with the lowest number of decimal places. This will tell you how many decimal places should be in the answer.

$1.7 + 0.812 + 10.30 = 12.812$, which rounds to 12.8 to 1 dp because 1.7 has the lowest number of decimal places (only 1 dp).

Multiplying and dividing: decide which number in the calculation has the least number of significant figures. This is the number of significant figures you should use in your answer.

$$\frac{17.70}{12.0} = 1.475 \quad 12.0 \text{ has 3 sf, while } 17.70 \text{ has 4 sf.}$$

So the answer is 1.48 to 3 sf.

Calculations with more than one step: many calculations involve more than one step, and if you use a calculator then you should use all the digits shown on the calculator at each stage and only round off at the end. Make sure that you read examination questions carefully because sometime they want you to quote a value to a certain number of significant figures before the final stage of the calculation.

Apply this skill by doing Activities A1, A4, A7, A8, A9, A10, A12, A15, A16, AIS1, AIS2.

S33 FINDING ARITHMETIC MEANS

Sometimes you may need to find the **mean** of a set of results. During your course this happens in **volumetric analysis**. In S7, Table 3 we have an example of two precise titres (volumes), 25.40 cm^3 and 25.45 cm^3. To find the mean, add the values and then divide by the number of values you have added.

$$\frac{25.40 + 25.45}{2} = 25.43 \text{ cm}^3 \text{ So the mean titre is 25.43 cm}^3.$$

Apply this knowledge by doing Activities A1, AIS1.

S34 CHANGING THE SUBJECT OF AN EQUATION

The equation we are referring to here is not a chemical equation but a mathematical expression, or formula.

At some stage in your course you will study enthalpy changes and these are given the symbol ΔH, where Δ means *change of* and H is the symbol for enthalpy.

One equation involving enthalpy changes could be:

$\Delta H_1 = \Delta H_2 + \Delta H_3$

Here, ΔH_1 is the **subject of the equation**, because it is on its own, usually on the left-hand side.

Suppose we want to make ΔH_2 the subject; we will need to rearrange the equation.

The key thing to remember is that *whatever you do to one side of the equation you must do to the other*. This keeps the equation in balance.

Step 1 To make ΔH_2 the subject we subtract ΔH_3 from both sides.

$\Delta H_1 - \Delta H_3 = \Delta H_2 + \Delta H_3 \, (-\Delta H_3)$

this will leave ΔH_2 on its own

$\Delta H_1 - \Delta H_3 = \Delta H_2$

Step 2 Swap sides so that ΔH_2 appears on its own on the left.

$\Delta H_2 - \Delta H_1 = \Delta H_3$

In this example we have subtracted the same quantity from both sides to keep the equation in balance. On other occasions you may need to add the same quantity to change the subject.

Often, when we change the subject of an equation we may need to multiply or divide both sides. This happens with a concept you will meet very early in your course chemistry, that of *amount*. **Amount** to a chemist has the unit *moles* (see S18).

$$\text{amount in moles} = \frac{\text{mass in grams}}{\text{mass of one mole (in grams)}}$$

In this case the solid line tells us that we are dividing *mass in grams* by the *mass of one mole*.

We may need to know the *mass in grams*, so we make this the subject of the equation.

Step 1 To make *mass in grams* the subject, we multiply both sides by *mass of one mole*.

$$\text{amount in moles} \times \text{mass of one mole} = \frac{\text{mass in grams}}{\text{mass of one mole}} \, (\times \text{mass of one mole})$$

This will leave *mass in grams* on its own.

amount in moles $\times$ mass of one mole = mass in grams

Step 2 Swap sides.

mass in grams = amount in moles $\times$ mass of one mole

In this example we have kept the equation in balance by multiplying both sides by the same quantity.

As with many skills, being methodical about the way the working of a calculation is set out will allow you and your examiner to understand the steps in your calculation.

Apply this skill by doing Activities A16, AIS2.

S35 USING LOGARITHMS

Here we ask the question "Why use logs at all?"

In Activity 3 we consider successive ionisation energies.

For a magnesium atom:

- removing the first mole of electrons only takes 740 kJ mol⁻¹ of energy
- removing the fifth mole of electrons takes 18 000 kJ mol⁻¹
- and removing the 11th mole of electrons requires 169 900 kJ mol⁻¹.

For numbers such as these, which range over several orders of magnitude, using logarithms makes the numbers more manageable and often patterns are easier to see.

Note: When we talk about *order of magnitude* we are usually referring to the powers of 10. For example, 10^1 to 10^2 is an increase of one order of magnitude.

740 is $\log_{10}$ 2.869

18 000 is $\log_{10}$ 4.255

169 900 is $\log_{10}$ 5.230

You can see a graph of $\log_{10}$ of successive ionisation energies of magnesium against the number of electrons removed in A3.

The pH scale is a logarithmic scale. See Table 4.

$pH = -\log_{10} [H^+(aq)]$. Note $[H^+(aq)]$ means concentration of H^+ ions in mol dm⁻³.

As you can see, aqueous hydrogen ion concentrations can differ by, say 100 million million.

If you look back at S23 you can use your calculator to convert $[H^+(aq)]$ to pH by using the pH equation.

Apply this skill by doing Activities A3, AIS2.

$[H^+(aq)]/$ mol dm⁻³	pH
10^{-1}	1
10^{-2}	2
10^{-3}	3
10^{-4}	4
10^{-5}	5
10^{-6}	6
10^{-7}	7
10^{-8}	8
10^{-9}	9
10^{-10}	10
10^{-11}	11
10^{-12}	12
10^{-13}	13
10^{-14}	14

△ Table 4 Comparing $[H^+(aq)]$ values and pH values

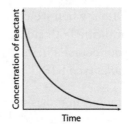

△ Fig 25 Rate at which a reactant is used up in a reaction.

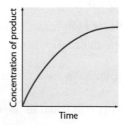

△ Fig 26 Rate at which a product is formed in a reaction.

S36 PLOTTING GRAPHS

If you are collecting **quantitative** data from an experiment to show the relationship between two variables, then plotting a graph is often the best way to look at the results. During your course you are most likely to be asked to plot results of rates of reaction experiments.

Rate of reaction is the change of concentration of a reactant, or product, per unit of time, as shown in Figs 25 and 26.

Let's look at this reaction, which is the hydrolysis of an ester:

$CH_3COOC_2H_5 + H_2O \rightarrow CH_3COOH + C_2H_5OH$
ester

We can determine the concentration of ester at different times and a graph of the results can be plotted (see Fig 27).

Graph plotting tips

- Look at the range of data you are going to plot on each axis to choose an appropriate scale.

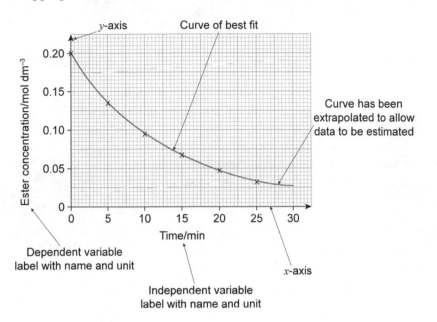

∧ Fig 27 The concentration of ester that remains in solution against time in the hydrolysis of an ester reaction.

- Your scale should allow the points you plot to take up more than half of the available space on each axis.
- The scale does not necessarily have to include zero, if zero is well away from the range of other points you are plotting.
- Make the scales so that it is easy to plot the data; this way you don't have to spend time calculating where you have to put the points.
- Use crosses because where they intersect is easy to see. Dots can sometimes be too faint, or too large.
- Rates of reaction graphs plot **continuous data**. This is data where each value can be any number between two limits. These are often straight lines or curves. A **line of best fit** goes through the points, so that most lie on the line or are roughly evenly spaced either side of the line.
- If a piece of data is obviously anomalous you should still plot it but show by a label or circle that it is anomalous; do not include it in a line of best fit.
- **Discrete data** is where the values are separate and can only be particular numbers. Graphs of ionisation energies (see A3 and A4) plot discrete data and each value is joined by a straight line. You do not draw a line of best fit because the data is discrete.

Apply this skill by doing Activities A3, AIS2.

If the graph is a straight line going through the origin (zeros) and it slopes up or down then this tells you that the two variables are **directly proportional**. This means that as one variable increases the other increases by the same percentage.

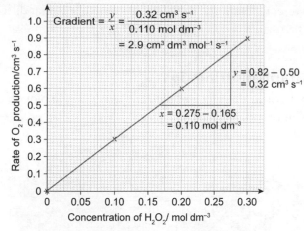

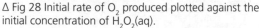

This relationship is represented by the equation:

$y = mx + c$

y is the y-axis value on the graph; x is the x-axis value

△ Fig 28 Initial rate of O_2 produced plotted against the initial concentration of H_2O_2(aq).

m is the slope (gradient), which for a straight-line graph is constant

c is the *intercept*, which is where the graph intercepts (hits) the y-axis.

In this case $c = 0$ so $y = mx + 0 = mx$

Fig 28 illustrates a method of obtaining rate vs concentration data by measuring initial rates for a number of different experiments. In some cases it is possible to get similar data from just one experiment. As a reaction proceeds it slows down as the concentration of the reactants decrease.

$$\text{Rate} = \frac{\text{change in concentration}}{\text{time}} = \text{slope of curve.}$$

We can see this by looking at the hydrolysis of ester reaction from S36 in Fig 29. At any point on the curve we can calculate the rate of reaction. To do this we draw a **tangent** (see Fig 30). A tangent is a line that just touches a circle at a point so the radius of the circle is perpendicular to that line. Fig 29 shows two tangents, one at time zero and the second at 20 min.

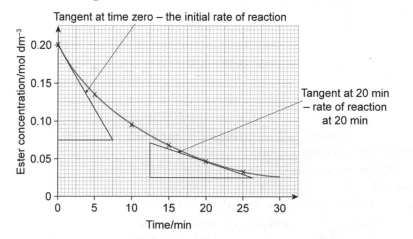

Tangent at time zero – the initial rate of reaction

Tangent at 20 min – rate of reaction at 20 min

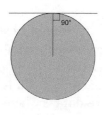

△ Fig 30 Tangent just touches the curve at one point.

△ Fig 29 The concentration of ester that remains in solution against time in the hydrolysis of an ester reaction.

Once we have drawn the tangent we find the slope in the same ways as for a straight-line graph.

$$\text{Slope at time zero} = \text{rate at time zero} = \frac{0.200 - 0.0750}{7.50 - 0} = \frac{0.125 \text{ mol dm}^{-3}}{7.50 \text{ min}}$$
$$= 1.67 \times 10^{-2} \text{ mol dm}^{-3} \text{ min}^{-1} \text{ to 3 sf}$$

The initial rate of reaction is the one at time zero. This is when the rate is fastest. We often use information from initial rates to determine orders of reaction. See S38 and A16.

The rate of reaction at 20 min $= \dfrac{0.070 - 0.0250}{26.5 - 12.5} = \dfrac{0.0450 \text{ mol dm}^{-3}}{14 \text{ min}}$

$= 3.21 \times 10^{-3} \text{ mol dm}^{-3}\text{min}^{-1}$

Calculating slopes is nothing to be afraid of: proceed methodically and include all the steps. Remember, even if you make a mistake you may receive credit from examiners if they can carry forward your error by recalculating your steps.

Apply this skill by doing Activity AIS2.

S38 ORDERS OF REACTION FROM GRAPHS

Rates of reaction are at their highest at the beginning of a reaction (see Fig 31). This is because the concentrations of reactants are at their highest values at the start. As the reaction proceeds reactants are used up, their solutions become less concentrated and the rate decreases. There is a relationship between rate of reaction and the concentration.

rate $\propto$ [reactant]^n

The square brackets mean *concentration of*.

$\propto$ means *proportional to*.

This can be expressed mathematically as:

rate $= k \times$ [reactant]^n or rate $=$ k[reactant]^n

This is called the **rate equation**.

- k is a constant called the *rate constant*.
- n is called the *order of reaction* with respect to a given reactant.

The rate equation can only be derived by experiment.

If $n = 1$ then the reaction is first order:

rate $= k$[reactant]

If $n = 2$ then the reaction is second order:

rate $= k$[reactant]^2

If $n = 0$ then the reaction is zero order and the rate is unaffected by the concentration of a particular reactant.

This can be shown graphically in Fig 31.

Reactions often involve more than one reactant. In A17 we use this rate equation:

rate $= k[NO(g)]^2[O_2(g)]$

Here the order of reaction with respect to NO is second order and with respect to O_2 it is first order. The overall order is third order; the sum of the individual orders.

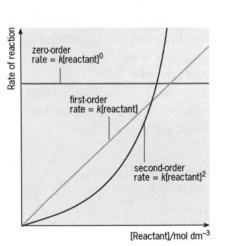

△ Fig 31 The effect of concentration on zero, first and second order reactions.

The rate equation is very important because it helps chemists work out the mechanism of a reaction, as you can see from A16.

Apply this skill by doing Activity A16, AIS2.

S39 APPRECIATING, VISUALISING AND REPRESENTING MOLECULES IN 2D AND 3D

Usually when you are asked to draw a displayed formula of methane, CH_4, you will draw the molecule shown in Fig 32.

It looks as if the bond angles are at right angles (90°) but this is not the case.

Methane has four bonding pairs of electrons around its central carbon atom. The bonding pairs repel each other so that they are as far apart as possible, giving bond angles of 109.5°. The shape of methane is tetrahedral because the hydrogen atoms are at the points of a tetrahedron, as you can see in Fig 33. This is predicted by the **electron pair repulsion model** and you can learn more about this in A6.

So that we can represent the methane molecule in 3D we use the way of representing the bonds shown in Fig 34.

All carbon atoms with four bonding pairs have a tetrahedral structure, as shown in Fig 35.

Δ Fig 32 Displayed formula of methane showing all atoms and bonds.

dot-and-cross diagram

the arrangement of bonds around the central carbon atom. All bond angles are 109.5°

hydrogen atoms are at the four corners of a tetrahedron

Δ Fig 33 Three ways to represent methane, a tetrahedral molecule.

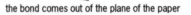

the bond comes out of the plane of the paper

the bond goes into the plane of the paper

the bond lies on the plane of the paper

Δ Fig 34 The three ways we represent the directions of bonds in a methane molecule.

Ethene has a double bond and in the electron pair repulsion model these electron pairs are one group of electrons and count as one bonding pair. This means we have three bonding pairs repelling each other, so the furthest apart they can get is 120°. This gives a trigonal planar shape round each carbon atom (see Fig 36). Trigonal tells us that the H atoms are where the points of a triangle would be and planar tells that the molecule lies flat on a plane, like the plane of your paper.

propane

Δ Fig 35 The three-dimensional shape of a propane molecule.

Δ Fig 36 Ethene is a trigonal planar molecule.

Representing molecules in 2D and 3D is an important skill to develop and becomes particularly important when studying E/Z isomerism and optical isomerism.

Apply this skill by doing Activities A6, A11, A12, A13, A14, A15.

Skills to Activities table

This table lists the activities that practise each skill.

Skill type	Skill	Links to activities
Working Scientifically	1 Explaining observations	A1, A2, A3, A4, A5, A6, A8, A9, A11, A12, AIS2
	2 Understanding how science advances	A1, A2, A3, A5, A6, A9, A10, A11
	3 Using appropriate techniques	A1, A6, A8, A9, A10, A11, A13, AIS1, AIS2
	4 Making valid observations	A1, A2, A5, A9, A11, AIS1, AIS2
	5 Taking measurements	A1, A2, A9, AIS1, AIS2
	6 Managing risk in investigations	A2, A10, AIS1
	7 Recording results appropriately	A3, A6, AIS1, AIS2
	8 Analysing and interpreting data	A1, A2, A3, A4, A5, A9, A11, A12, A13, A14, A16, AIS1, AIS2
	9 Identifying anomalous data	A1, A2, A4, A9, A11, A12, AIS1, AIS2
	10 Evaluating methodology, data and evidence	A1, A2, A5, A9, A11, A12, A14, AIS1, AIS2
	11 Assessing error	A1, A2, AIS1, AIS2
	12 Drawing valid conclusions	A1, A2, A5, A9, A11, A12, A14, A16, AIS2
	13 Communicating scientific information	A1, A5, A9, A10, A14
	14 Considering ethical issues	A7, A8, A10, A14
	15 Science and decision-making	A7, A8, A10, A14
	16 Writing balanced equations	A4, A8, A12, A16, AIS1
	17 Relative atomic mass	A4, A8, A12, A16, AIS1
	18 Amounts and the mole	A4, A8, A10, A11, A12, A14, AIS1, AIS2
	19 Relative molecular masses	A1, A7, A8, A10, A12, A13, A14, AIS1, AIS2
	20 Calculating amounts and masses from chemical equations	A8, A9, A10, A12, AIS1, AIS2
	21 Molar gas volumes	A8, A12, AIS2
	22 Calculating volumes of gases from chemical equations	A8, A12, AIS2
	23 Working with amounts in aqueous solutions	AIS1, AIS2
	24 Titrations	AIS 1, AIS2

Skill type	Skill	Links to activities
Quality of Written Communication	**25** Writing for your intended audience	**A1–A16**
	26 Ensuring meaning is clear	**A1–A16**
	27 Organising information clearly and coherently	**A1–A16**
	28 Using specialist vocabulary	**A1–A16**
Maths	**29** Standard form	**A1, A8, A9, A11, A12, A15, A16**
	30 Ratios, fractions and percentages	**A1, A8, A9, A11, A12, A15, A16**
	31 Using your calculator	**A1, A3, A4, A7, A8, A9, A10, A11, A12, A13, A14, A15, A16, AIS1, AIS2**
	32 Using an appropriate number of significant figures	**A1, A4, A7, A8, A9, A10, A12, A15, A16, AIS1, AIS2**
	33 Finding arithmetic means	**A1, AIS1**
	34 Changing the subject of an equation	**A16, AIS2**
	35 Using logarithms	**A3, AIS2**
	36 Plotting graphs	**A3, AIS2**
	37 Determining slopes and working out rates of reaction	**AIS2**
	38 Orders of reaction from graphs	**AIS2**
	39 Appreciating, visualising and representing molecules in 2D and 3D	**A6, A11, A12, A13, A14, A15**

Activities

A1 IRON-60 FOUND ON EARTH

Elements that are heavier than hydrogen and helium are produced in stars by nuclear fusion. A star's life can sometimes end in a stupendous explosion, called a supernova, which flings out the elements created together with cosmic rays.

△ Fig 37 Supernova explode, spewing out material at one tenth the speed of light.

In the 1950s a scientist suggested that a nearby supernova might have caused a mass extinction of marine creatures two million years ago. Then, in 1996, scientists in the USA proposed that if radioactive atoms were to be found in fossils this could be evidence of nearby supernovae explosions. Two years later a German team found traces of iron-60 deep in the Pacific Ocean floor, in rock samples formed two million years ago. Iron-60 is unstable, with a half-life of 2.62 million years, which means that none should exist naturally on Earth because it would have decayed into other elements long ago. But if a more recent supernova had occurred then traces of iron-60 might be detected.

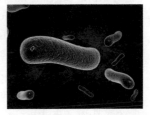

△ Fig 38 Bacteria deep in the ocean convert iron into tiny crystals of magnetite, no larger than 80 nanometres in diameter.

An astrobiologist suggested that the team try looking at rocks with magnetite crystals in them. These minute crystals are produced by bacteria from iron in atmospheric dust that falls into the oceans. When these bacteria die they form a fossil imprint in sediment laid down on the ocean bed. Sure enough, in 2004 they found evidence of iron-60 in a layer of magnetite crystals, followed in 2013 by yet more evidence sparking worldwide interest.

QUESTIONS

1. Magnetite crystals are approximately 80 nanometres in diameter. Use standard form to write this diameter in metres.

2. From analysis, magnetite has a relative formula mass of 159.6 and a percentage composition by mass of Fe, 69.9%; O, 30.1%. What is the formula of this iron oxide? (A_r: Fe = 55.9; O = 16.0)

3. The abundance of different iron isotopes found on Earth are shown in Table 5.

	Iron-54	Iron-56	Iron-57	Iron-58
Abundance %	5.84	91.68	2.17	0.31

△ Table 5

Calculate the relative atomic mass of iron to two decimal places.

4. a) The German scientists announced their discovery in a letter to a respected scientific journal. It took nine months for it to be published. Explain why there was such a delay.
   b) The German team now intend to analyse an even larger sample of rock. Why is it important that they do this?

5. Explain why this story illustrates how scientists are working scientifically. (Hint: Look at Skill 13.) [QWC]

Skills practised
1, 2, 3, 4, 5, 8, 9, 10, 11, 12, 13, 17, 19, 25, 26, 27, 28, 29, 30, 31, 32, 33

A2 THE NUCLEAR MODEL OF THE ATOM

The electron was the first subatomic particle to be discovered. It was identified by J. J. Thomson in 1897. It was negatively charged and he calculated its mass to be about two thousand times less than a hydrogen atom. Thomson's model of the atom had thousands of electrons making up the atomic mass, embedded in a cloud of positive charge with no mass.

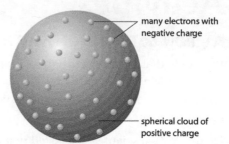

△ Fig 39 Thomson's model of the atom. The negative charge of the electrons is cancelled out by the positively charged sphere.

At about the same time the phenomenon of radioactivity was discovered. One of Thomson's team, Ernest Rutherford, suggested that there were different types of radiation. He found evidence for two of these and called them alpha particles and beta particles. He worked out the mass of an alpha particle to be the same as a helium atom but with a positive charge. This suggested to him that an atom might have positive subatomic particles.

In 1912, Rutherford went on to test Thomson's model by firing alpha particles at very thin gold foil, just a few atoms thick. He expected the alpha particles to crash through the foil, with slight deflections, because there was nothing much inside Thomson's atom to stop them. To his astonishment one in every ten thousand was reflected back.

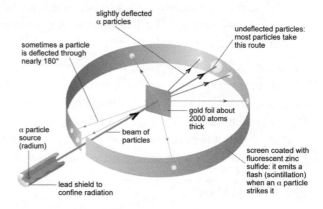

△ Fig 40 Rutherford's scattering experiment.

From these results Rutherford proposed his model of the atom with a tiny, positive, central nucleus, where almost all the mass was concentrated, surrounded by empty space. The electrons orbited the nucleus, kept apart by their repelling negative charge.

QUESTIONS

1. Identify a major hazard in Rutherford's scattering experiment and suggest what could be done to reduce the risk from this hazard.

2. Which results are anomalous?

3. Why did Rutherford insist that the experiment was repeated many times?

4. What was the experimental evidence for:

   a) A very tiny nucleus surrounded by lots of empty space?
   b) A nucleus where most of the mass of the atom is concentrated?
   c) A positively charged nucleus?

5. Use the development of Rutherford's nuclear model to show how science advances and how observations are explained scientifically. **QWC**

Skills practised

1, 2, 4, 5, 6, 8, 9, 10, 11, 12, 25, 26, 27, 28

A3 THE ARRANGEMENT OF ELECTRONS IN SHELLS

In the 1920s new models of the atom were proposed, still with the tiny nucleus of Rutherford's model but with the electrons arranged in energy levels, or shells (see Fig 41).

Evidence for the existence of shells comes from successive ionisation energies (IE). The first ionisation energy is the energy required to remove 1 mole of electrons from 1 mole of gaseous atoms: $E(g) \rightarrow E^+(g) + e^-$ (E is any element; (g) means gaseous). The second ionisation energy is the energy to remove a second mole of electrons: $E^+(g) \rightarrow E^{2+}(g) + e^-$

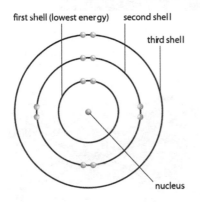

△ Fig 41 The arrangement of electrons in shells of a magnesium atom.

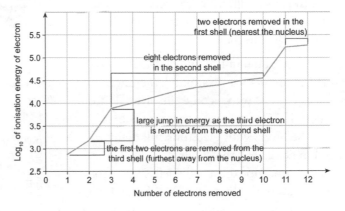

△ Fig 42 Graph of $\log_{10}$ of successive IEs of magnesium against the number of electrons removed. It does not require much extra energy to be used to remove electrons from the third shell.

QUESTIONS

1. Give the equation for the 12th ionisation energy of magnesium.

2. Table 6 shows successive ionisation energies for sodium.

Number of electrons removed/ mol	1st	2nd	3rd	4th	5th	6th	7th	8th	9th	10th	11th
IE/kJ mol^{-1}	500	4600	6900	9500	13 400	16 600	20 100	25 500	28 900	141 000	159 000

△ Table 6 Successive ionisation energies for sodium in kJ mol^{-1} of electrons removed.

   a) Draw a suitable table and plot a graph of $\log_{10}$ of successive IE against number of moles of electrons removed. Calculate $\log_{10}$ values to 2 significant figures.
   b) Why is it sensible to convert ionisation energies to $\log_{10}$ in this graph?
   c) Explain how your graph provides evidence for the existence of shells.

3. The first two ionisation energies of caesium and barium are shown in Table 7. Explain this data in terms of the electron configuration of these atoms. **QWC**

Element	First IE/kJ mol^{-1}	Second IE/kJ mol^{-1}
Caesium	376	2234
Barium	503	965

△ Table 7

Skills practised

1, 2, 7, 8, 25, 26, 27, 28, 31, 35, 36

A4 THE ARRANGEMENT OF ELECTRONS IN SUBSHELLS

The spectrum of visible light is continuous, which means all frequencies are represented. When atoms are given extra energy (excited) they release this energy but only certain frequencies are emitted. These show up as a series of separate lines called the emission spectrum of the element. Different elements have different emission spectra with lines of different frequencies. These observations led to the model of an atom with electrons arranged in shells. (See A3.)

When the fine detail of the line emission spectra of elements was examined, the bright lines (as in Fig 43) were seen to be divided into more lines. This led to a refinement of the model of the atom, with electrons still being arranged in shells but the shells now divided into subshells. Names for these subshells come from the finer detail of the line spectrum. Some lines were sharp, hence *s* subshells, some were brighter and called *principal* lines, hence *p* subshells and some were spread out or *diffuse*, hence *d* subshells.

Further evidence for subshells comes from plotting the first ionisation energies of successive elements.

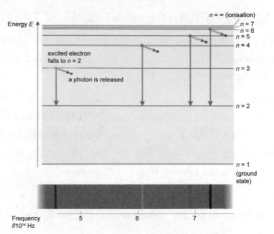

△ Fig 43 Atomic emission spectra are seen when energy is given to atoms. This is the line emission spectrum of hydrogen.

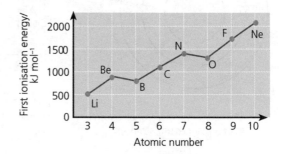

△ Fig 44 First ionisation energies across Period 2.

QUESTIONS

1. The first ionisation energy of lithium is 520.2 kJ mol⁻¹.
   a) Give the equation for this ionisation energy.
   b) Work out the energy required to remove the outermost electron from one atom of lithium. Your answer should be quoted to the appropriate number of significant figures. (Hint: You need to know the number of atoms in 1 mole.)

2. Explain the general trend for first ionisations across the period.

3. Table 8 gives the first ionisation energies of successive elements in Period 3.

Element	Na	Mg	Al	Si	P	S	Cl	Ar
IE/kJ mol⁻¹	496	738	578	787	1040	1000	1251	1521

△ Table 8

   a) Draw a graph similar to Fig 44, using the information from Table 8.
   b) Compare your graph with Fig 44. What similar pattern is present in both graphs?

4. Periods 2 and 3 each have an s subshell containing up to two electrons and a p subshell containing up to six electrons. Using Fig 44, explain:
   a) the first ionisation energy increases from Li to Be.
   b) the first ionisation energy decreases from Be to B.
   c) the first ionisation energy decreases from N to O.
   (Hint: Once there are three electrons in the p subshell they start to pair up.) **QWC**

Skills practised

1, 8, 9, 18, 25, 26, 27, 28, 31, 32, 36

A5 NOBLE GASES REACT!

In 1916 two American chemists, Gilbert Lewis and Irving Langmuir, realised independently that all the noble gases except helium had eight electrons in their outer shells and suggested it was this electron configuration that made noble gases unreactive. They put forward the idea that when atoms of other elements reacted to form compounds they gained, lost or shared electrons to make a noble gas outer shell of eight electrons – or two for elements close to helium. The noble gases were thought to be inert (totally unreactive and so unable to form compounds) and Group 0 elements were even given the name "the inert gases".

Lewis went on to propose a model for ionic and covalent bonding based on what we now call dot and cross diagrams. Although this model is very simple it is still very useful today and it paved the way for huge advances in our understanding of chemical reactions.

However, in 1962 a British chemist, Neil Bartlett, succeeded in synthesising xenon hexafluoroplatinate, $Xe^+[PtF_6]^-$, the first Group 0 compound to be made. Bartlett predicted that the platinum compound, PtF_6, might remove an electron from xenon because, by accident, he had produced another platinum compound, $O_2^+PtF_6^-$. This involved removing an electron from the $O_2(g)$ molecule, making PtF_6 the most powerful oxidising agent yet discovered. He knew that the energy required to do this was similar to the first ionisation energy of xenon. (See Activity 4.)

△ Fig 45 Neil Bartlett holding the apparatus he used to make the first Group 0 compound. It formed in the glass container at the bottom of the apparatus.

He set up a simple experiment with xenon gas in one glass container and PtF_6 in the other, removed the seal between the containers and allowed the two gases to mix. A yellow solid immediately formed.

He attempted to get his discovery published in *Nature*, a respected scientific journal, but it was slow to respond. After two months he withdrew his report from *Nature* and sent it to *Proceedings*, another distinguished journal. They published it the next month and it caused a sensation, with chemists rushing to make more noble gas compounds. Three years later there were whole textbooks dedicated to noble gas chemistry. Bartlett himself went on to make XeF_2, XeF_4 and XeF_6.

QUESTIONS

1. How do ionisation energies provide evidence for the stability of noble gases? (Hint: Look at Fig 45 in Activity 4.)
2. Showing only the outer shell electrons, use dot and cross diagrams to draw an atom of sodium, an atom of chlorine and the ions of sodium chloride.
3. Draw dot and cross diagrams of CH_4, NH_3, H_2O.
4. Why do the dot and cross diagrams of BF_3 and SF_6 not fit the 1916 ideas of Lewis and Langmuir?
5. Explain why:
   a) the journal *Nature* might have been slow to respond
   b) Bartlett felt it was essential to get his findings published quickly.

Skills practised:
1, 2, 4, 8, 10, 12, 13, 25, 26, 27, 28

A6 THE SEARCH FOR NEW MEDICINES

Drugs work largely because of their shape. They act on complex receptor sites within the body rather like a key fitting into a lock. A technique called X-ray crystallography can be used to work out the shape and structure of these receptor sites. Once the shape is known, work can begin on designing a drug to fit it. Just because a molecule fits the site it does not mean it will have any activity that will be beneficial to treating an illness, but it is a start.

It is very important that chemists can predict molecular shapes and the basis for these predictions is the electron pair repulsion model. A pair of electrons in a covalent bond is called a **bonding pair**. Pairs of electrons that are not involved in bonding are called **lone pairs** (**non-bonding pairs**). In this model, pairs of electrons around the central atom repel each other and move as far apart from each other as possible.

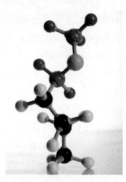

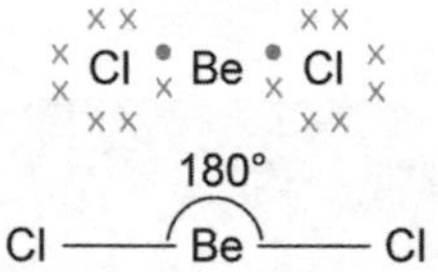

△ Fig 46 The bond angles of this molecule can be predicted using the electron pair repulsion model.

$$\times\!\times \qquad \times\!\times$$
$$\times\ Cl\ {}^{\bullet}_{\times}\ Be\ {}^{\bullet}_{\times}\ Cl\ \times$$
$$\times\!\times \qquad \times\!\times$$

Cl ——— Be ——— Cl (180°)

△ Fig 47 Beryllium chloride, BeCL$_2$, has two bonding pairs of electrons around the central atom, so its shape is linear.

QUESTIONS

1. There is a model of drug action given above. What name could be given to this and why is this model useful?

2. The following molecules can be used to illustrate how the electron pair repulsion model predicts shapes: from Fig 47 BeCl$_2$ is linear (bond angles of 180°); BCl$_3$ is trigonal planar (bond angles of 120°); CH$_4$ is tetrahedral (bond angles of 109.5°).

   Draw a table that contains this information and include dot and cross diagrams of the molecules and drawings that clearly show the arrangement of bonds around the central atom.

3. Predict, with reasons, the bond angles of the atoms surrounding the blue atom in Fig 46.

4. The bond angle of NH$_3$ is 107° and that for H$_2$O is 104.5°. Draw dot and cross diagrams and show the arrangement of the bonds in these molecules. What does this indicate about the relative strength of repulsion of lone pairs compared with bonding pairs?

5. Computers can be used to design possible drug molecules on screen. Different groups of atoms can be made to probe the active site because molecular shape is easily visualised using computer graphics. Previously this was all done in the laboratory. Explain why using computer graphics to begin the design of drug molecules is now used extensively by all pharmaceutical companies. QWC

Skills practised
14, 15, 19, 25, 26, 27, 28, 30, 31, 32

A7 BIOETHANOL AND BRAZIL

Brazil has been growing sugar cane for centuries and now almost all of the juice extracted goes into the production of ethanol for fuel. The juice is first fermented then distilled.

In the 1970s, two oil crises caused oil prices to rise dramatically. At this time Brazil was heavily dependent on imported oil, which it could no longer afford. This provided the stimulus for the government to start its Proalcool programme to reduce fossil fuel dependency. The government decreed that all car engines made after 1979 had to run on ethanol, or a blend of petrol (gasoline) with 22% ethanol. By 1984 more than 80% of cars that were manufactured in Brazil had 'ethanol engines'.

In 1986 the effects of the oil crises eased and by 2001 only 1% of cars being manufactured were ethanol-only engines, although the Brazilian government still insisted on the blended ethanol-gasoline mix. The popularity of bioethanol has increased in Brazil since 2003 due to the development of a flexible-fuel vehicle that could run on a mixture containing any proportion of petrol to alcohol.

△ Fig 48 Some of Brazil's sugar cane is still harvested by hand.

△ Fig 49 Ethanol became a very important car fuel in Brazil in the 1980s.

QUESTIONS

1. Ethanol produced from sugar cane is a *biofuel* and may be referred to as *carbon-neutral*. What is meant by these terms?

2. In terms of the carbon dioxide balance in the atmosphere how does burning sugar for fuel compare with:
   a) burning oil for fuel?
   b) producing sugar for food?

3. Energy density is the energy released from burning 1 kg of fuel. It can be calculated from the enthalpy change of combustion of the fuel and its M_r.
   a) Calculate the energy density of ethanol to 3 significant figures. ($\Delta H_c^{\ominus} = 1370$ kJ mol^{-1})
   b) i) The energy density of petrol is 55% greater than that of ethanol. Work out the energy density of petrol.
      ii) 1 litre of petrol has a mass of 740 g. How many litres of petrol are there in 1 kg?
      iii) The price of ethanol at the fuel pump fluctuates. Flexible-fuelled vehicles can switch to using only ethanol. How cheap, in terms of a percentage of the price of petrol, would bioethanol have to be before it would be advisable for a motorist to switch? (Assume 1 kg of ethanol has the same volume as 1 kg of petrol.)

4. Some scientists propose that we use more biofuels, while others are concerned about the vast areas of land that are required to grow the necessary crops. Apart from carbon-neutrality, what are some of the advantages and disadvantages of producing more biofuel? **QWC**

Skills practised

14, 15, 19, 25, 26, 27, 28, 30, 31, 32

A8 HYDROGEN: CAR FUEL OF THE FUTURE?

One major disadvantage of using hydrogen as a car fuel is that it is difficult to store. One way would be to liquefy it but this requires energy and even then hydrogen has to be kept below −253 °C. Another possible method is to store it in a solid form – and this does not mean freezing it! Metal hydrides are compounds containing hydrogen; one candidate is magnesium hydride MgH_2. A litre of magnesium hydride contains almost as much hydrogen as a litre of liquefied hydrogen, although it is much heavier.

The first carbon nanotube was prepared in Japan by Iijima Sumio in 1991, spurred on by the discovery of the first fullerene (see A9). This single-walled tube is like a layer of graphite wound into a cylinder with a diameter ranging from 1 to 50 nm. Many uses have been suggested for nanotubes, one of which is for the on-board storage of hydrogen, where hydrogen is adsorbed onto the nanotube walls at only slightly increased pressure.

△ Fig 50 A single nanotube.

At first the uptake of hydrogen by nanotubes was only modest, but then researchers in Germany began to design a material based on the network found in natural sponges. They used advanced mathematics and computer models to suggest that the uptake could be greatly increased by using a network of carbon nanotubes. The combined weight of the nanotubes and the hydrogen adsorbed is predicted to be 5.5% of the total weight of the vehicle and the material would be cheap, non-toxic and lightweight.

Several groups of researchers are trying to make these sponge-like networks of nanotubes but, to date, no one has succeeded. The German team are also setting about making the material experimentally.

QUESTIONS

1. What is the percentage by mass of hydrogen in magnesium hydride? (A_r: Mg = 24.3; H = 1.0)

2. Hydrogen is released by the reaction of magnesium hydride with water to produce magnesium hydroxide.
   a) Construct an equation for this reaction and include state symbols.
   b) Calculate the volume of hydrogen produced per kilogram of magnesium hydride at 298 K and 101 kPa.

3. When hydrogen is adsorbed, what intermolecular force is likely to hold the hydrogen on the walls of the nanotubes? Explain your reasoning.

4. Hydrogen is easily released from the nanotube matrix inside the fuel tank. As the engine takes in hydrogen, the pressure in the tank falls, which releases more hydrogen. Use your understanding of dynamic equilibria to explain why more hydrogen is released.

5. What are the advantages and disadvantages of switching from petrol to hydrogen as a fuel? **QWC**

Skills practised
1, 3, 14, 15, 16, 18, 19, 20, 21, 22, 25, 26, 27, 28, 29, 30, 31, 32

A9 FULLERENES – NEW FORMS OF CARBON

In the 1970s Harry Kroto, a chemist from Sussex University, together with two Canadian astronomers, detected long chains of carbon atoms in interstellar space. Harry's hypothesis was that they might have formed in the carbon-rich clouds of red giant stars. He had the opportunity to test his idea when he met two American scientists, who were working with a laser machine that fired pulses reaching temperatures far higher than in most stars. In 1985, they vapourised graphite in an atmosphere of helium and analysed the resulting carbon species using a mass spectrometer.

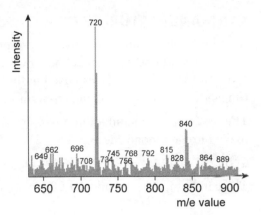

△ Fig 51 The mass spectrum of laser-vaporised graphite showing a large peak at 720 m/e.

The experiment did find evidence for the formation of long-chain carbon molecules but it also produced a very strange result: a large peak at 720 m/e on the mass spectrum. This corresponded to a C_{60} molecule. They worked out that the structure for this species was similar to the geodesic domes designed by the architect Robert Buckminster Fuller. Hence the name by which we now know this new class of molecules – fullerenes.

A letter to the scientific journal *Nature* in 1985, announcing their discovery, produced a sceptical reaction from many scientists. A peak in the mass spectrum was not considered enough evidence for a completely new form of carbon, largely because the amount to produce this peak is extremely small. Then in 1990 Wolfgang Kratschmer announced that he had prepared C_{60} by vapourising graphite in an inert atmosphere in a sufficient amount to allow him to analyse it. The structure was indeed the cage structure shown in Fig 52.

△ Fig 52 Buckminsterfullerene, C_{60}, the first fullerene to be discovered.

QUESTIONS

1. Look at Fig 51.
   a) Why does the large peak suggest a species with the formula, C_{60}?
   b) The C_{60} species produced in the mass spectrometer is not a molecule. What is it and how is it formed?
   c) What species does the peak at 840 suggest?
   d) Why is the species at 720 considered to be stable?

2. a) Why did Kratschmer use an inert atmosphere to vaporise graphite rather than just air?
   b) Draw the structures of graphite and diamond. Use these to suggest a reason why Kratschmer did not use diamond.

3. In an experiment, 1.208 grams of graphite is vaporised to give pure C_{60} of 0.0306 milligrams.
   a) Express 0.0306 milligrams in grams using standard form.
   b) Calculate the percentage yield of C_{60}.

4. Explain why Kratschmer's experiment convinced scientists that C_{60} had indeed been produced by Kroto and his colleagues. **QWC**

> **Skills practised**
> 1, 2, 3, 4, 5, 8, 9, 10, 12, 13, 17, 20, 25, 26, 27, 28, 29, 30, 31, 32

A10 MANUFACTURING NITRIC ACID – A GREENER WAY

About 60 million tonnes of nitric acid is produced globally each year. It is a very important chemical and almost 80% goes into making ammonium nitrate (NH_4NO_3) fertiliser. It is also used in the manufacture of nylon, dyes, fungicides, explosives and some pharmaceuticals.

The first stage of manufacture involves oxidising ammonia in excess air to make nitrogen monoxide (see Fig 53):

$$4NH_3(g) + 5O_2(g) \rightarrow 4NO(g) + 6H_2O(g) \; \Delta H° = -909 \text{ kJ mol}^{-1}$$

The catalyst used is an alloy of 90% platinum and 10% rhodium.

Δ Fig 53 A mesh of platinum and rhodium catalyses the oxidation of ammonia.

However, it is also at this stage that other reactions occur that produce nitrous oxide (N_2O). One of these reactions is:

$$4NH_3(g) + 4O_2(g) \rightarrow 2N_2O + 6H_2O(g)$$

Nitrous oxide is a very potent greenhouse gas. It has a global warming potential 310 times that of carbon dioxide, so every tonne of nitrous oxide has the same global warming effect as 310 tonnes of carbon dioxide. This is largely because it remains in the atmosphere for more than a hundred years without decomposing. There is great concern over the production of nitrous oxide as a by-product of nitric acid manufacture and recent legislation has set limits on how much nitrous oxide can be produced for every tonne of nitric acid. This has stimulated research into new catalysts and modifying existing plant to try and remove most of the nitrous oxide produced.

QUESTIONS

1. Ammonium nitrate is a very important fertiliser because it contains a high percentage composition of nitrogen, an element essential for crop production. Calculate the percentage composition of nitrogen in this compound.

2. A nitric acid plant produces 9.24 tonnes of nitrogen monoxide for every 320 000 moles of ammonia used. Calculate the percentage yield of nitrogen monoxide.

3. In the first stage of the manufacture of nitric acid the reaction could reach dynamic equilibrium. Explain what is meant by the term *dynamic equilibrium*.

4. In practice, the manufacturers do not wait for dynamic equilibrium to form and the conditions used are 900 °C and 1000 kPa.
   a) Explain why a manufacturer does not wait for dynamic equilibrium to form and why these conditions are used.
   b) Use Boltzmann distribution diagrams to explain the effect of increasing the temperature on the rate of reaction.

5. a) Why is the catalyst in the form of a mesh? (See Fig 53.)
   b) What effect does the catalyst have on the position of equilibrium?
   c) Explain, using a Boltzmann distribution diagram, the effect of using a catalyst on the reaction rate.

6. a) Explain how atmospheric nitrous oxide molecules contribute to global warming.
   b) State two modern analytical techniques that scientists can use to monitor the production of nitrous oxide from a nitric acid plant.
   c) Why might legislation stimulate research into reducing nitrous oxide levels from the manufacture of nitric acid?

7. Why is it important to establish international cooperation to reduce emission levels of nitrous oxide?

8. Explain the conditions that should be chosen to maximise the percentage yield of nitric oxide in this dynamic equilibrium reaction. Remember that these are not the same as the conditions used by the manufacturer in Question 4. **QWC**

Skills practised

2, 3, 6, 13, 14, 15, 19, 20, 21, 25, 26, 27, 28, 31, 32

A11 A STRUCTURAL MODEL FOR BENZENE

When benzene was first isolated chemists could not understand how its properties could be explained by its structure. A major step in understanding benzene's structure was taken in 1865 by August Kekulé, a German chemist. He proposed that benzene had a ring of carbon atoms with alternating double and single bonds. This was the first time chemists began to accept that carbon could form rings as well as chains.

There were, however, faults with Kekulé's model. For a start, as modern analytical techniques were developed, X-ray crystallography and infrared spectroscopy indicated a molecule where all six C–C bonds were the same length at 140 pm. This was in between the length of a C–C single bond and a C=C double bond.

The modern model of benzene was developed during the 20th century. In this model, when a hydrogen atom forms a bond with a carbon atom a p-orbital is perpendicular to the plane of the carbon ring. Each p-orbital contains a single electron and the orbitals overlap to form a ring above and below the carbon ring. The six electrons are not found between any particular carbon atoms so they are said to be delocalised π electrons forming π bonds.

△ Fig 54 A snake biting its tail in a dream was the inspiration for Kekulé's structure of benzene.

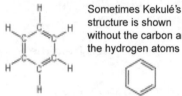

Sometimes Kekulé's structure is shown without the carbon and the hydrogen atoms

△ Fig 55 Kekulé's structure of benzene.

overlap

n bonds are formed by side-by-side overlap of all six 2p atomic orbitals

Fig 56 π bonds are formed by the sideways overlap of the six 2p orbitals.

QUESTIONS

1. Express 140 pm in metres using standard form. (A picometre is 10^{-12} metres.)

2. The enthalpy change of hydrogenation is the enthalpy change when one mole of alkene reacts with hydrogen to form an alkane.

   a) The enthalpy change of hydrogenation of cyclohexene (Fig 58) is -120 kJ mol^{-1}. Give an equation for this reaction.
   b) Predict the value of the enthalpy change of hydrogenation of Kekulé's structure of benzene. (Hint: Consider the number of C=C bonds.)

3. The experimental result for the enthalpy change of hydrogenation of benzene is -208 kJ mol^{-1}.

   a) Draw energy level diagrams of the predicted enthalpy change of hydrogenation of Kekulé's benzene from 2.b) above and the actual experimental value for the enthalpy change.
   b) Explain how this shows that the benzene molecule is more stable than predicted.

4. Use the development of the modern model of benzene to explain how science advances. **QWC**

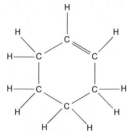

△ Fig 57 Cyclohexene.

Skills practised

1, 2, 3, 4, 8, 9, 10, 12, 19, 25, 26, 27, 28, 29, 31, 39

A12 TNT – A FORMIDABLE EXPLOSIVE

It is remarkable to think that when TNT was first made in 1863 it was used as a yellow dye. It was some years before its potential as an explosive was realised. This is because it has a very high activation energy and does not explode under normal conditions. Although not a powerful explosive compared with some other explosives discovered at the same time, its very stability makes it safer to use in blasting out quarries and building roads and railways. However, a major use still remains as an explosive in artillery shells and mines and it was used extensively during the two world wars.

△ Fig 58 TNT is being used to blast this limestone quarry.

Understanding the mechanism of the reaction to make TNT helps in understanding why Kekulé's model of benzene in A11 was incorrect. In this model benzene has three carbon-carbon double bonds and we would expect it to react like alkenes in easily undergoing electrophilic addition reactions with reagents like bromine water. In fact benzene will not react with bromine water. It will react with liquid bromine and a catalyst but even then it does not react in an addition reaction but in an electrophilic substitution reaction. Benzene rings hardly ever undergo addition reactions.

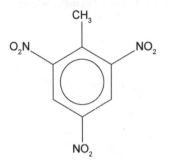

△ Fig 59 TNT has an IUPAC name 2,4,6-trinitromethylbenzene.

QUESTIONS

1. Draw the structure of methylbenzene.

2. An explosion is an extremely rapid expansion of hot gases from a small volume of explosive, in this case solid TNT.
   a) When TNT first explodes there is no time for it to react with oxygen, it just decomposes. The products are nitrogen, water, carbon monoxide and carbon. Construct a balanced equation for this reaction.
   b) Use your equation to calculate the total volume of gases produced at 298 K and 101 kPa per mole of TNT. Remember that H_2O will be a gas when TNT explodes.
   c) What volume of gases is produced from 1 gram of TNT?
   d) An explosion of TNT is accompanied by lots of black smoke. What is the explanation for this?

3. What is meant by the term *activation energy*?

4. a) If Kekulé's structure (A11) is correct, it should undergo reaction by electrophilic addition with bromine water. Draw the structure of the product that would be predicted to form.
   b) In reality bromine reacts via the electrophilic substitution mechanism to produce bromobenzene (C_6H_5Br). Write a balanced equation for this reaction.
5. Draw the curly arrow mechanism, to show the mononitration of methylbenzene to produce 4-nitromethylbenzene. Your answer should identify the electrophile and explain the meaning of this term. It should also include the reagents and conditions. **QWC**

Skills practised
1, 8, 9, 10, 12, 16, 18, 19, 20, 21, 22, 25, 26, 27, 28, 29, 31, 31, 39

A13 PEPPERMINT IN MEDICINE

Peppermint leaves have been used as a medicinal herb for centuries, showing beneficial effects in treating nausea, indigestion and disorders of the bowel. Peppermint oil can also be extracted from the peppermint plant and it too is being investigated for the treatment of a very painful and unpleasant illness called irritable bowel syndrome and for its potential protective effect in radiological cancer treatments.

Analysis of peppermint oil, using gas chromatography (GC) followed by mass spectrometry (MS), has shown that two chemicals present in high concentrations are menthol (43%) and menthone (24%).

△ Fig 60 Peppermint oil and the leaves of the peppermint plant.

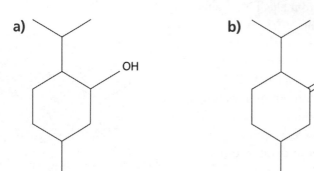

△ Fig 61 The skeletal formula of a) menthol and b) menthone.

QUESTIONS

1. a) Explain what is meant by the term functional group.

    b) Identify the functional group in each of these compounds.

2. Calculate the relative molecular masses of each compound.

3. a) In the laboratory 2,4-dinitrophenylhydrazine (DNPH) can be used to distinguish between these two compounds. Describe the results you would expect for each compound.

    b) Explain why Tollens' reagent cannot be used to distinguish between these two compounds.

4. a) Describe the reagents and conditions you could use to prepare menthone from menthol.

    b) What reagent could convert menthone into menthol?

5. Explain why combining gas chromatography with mass spectrometry makes a very powerful analytical tool.

6. Menthol and menthone show optical isomerism. Explain what is meant by this term using one of these compounds as an example. Your answer should indicate the chiral carbons in each molecule. **QWC**

> **Skills practised**
> 3, 8, 19, 25, 26, 27, 28, 31, 39

A14 SATURATED AND UNSATURATED FATTY ACIDS

The only difference between a fat and an oil is its physical state at room temperature: fats are solids and oils are liquids.

Fats and oils are esters; esters are the product of a reaction between a carboxylic acid and an alcohol (see Fig 62). The alcohol involved has the trivial (common) name of glycerol.

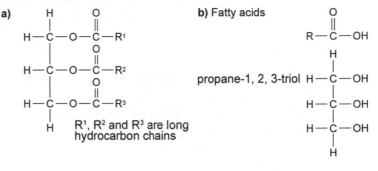

△ Fig 62 The structure of (a) fats and oils and (b) the alcohol and fatty acids that react to make them.

The naturally occurring carboxylic acids that form the esters are often referred to as fatty acids. In Table 9 three fatty acids are shown. The trivial names come from their sources, so for example oleic acid from olive oil, even though oleic acid is found in many plant oils and animal fats.

Trivial name	Systematic name	Structural formula
Stearic acid	octadecanoic acid,18,0	$CH_3(CH_2)_{16}COOH$
Oleic acid	octadec-9-enoic acid, 18,1(9)	$CH_3(CH_2)_7CH=CHCH_2(CH_2)_6COOH$
Linoleic acid	octadec-9,12-dienoic acid, 18,2(9,12)	$CH_3(CH_2)_4(CH=CHCH_2)_2(CH_2)_6COOH$

△ Table 9

Vegetable oils tend to have a greater proportion of unsaturated fatty acids than animal fats. This means that the fatty acids contain at least one double bond. Saturated fats are linked to the production of high levels of cholesterol in the blood, which increases the risk of heart disease and strokes.

QUESTIONS

1. **a)** What is the systematic (IUPAC) name for glycerol?

    b) Name the two functional groups present in oleic acid.

2. **a)** Why are systematic names preferred to trivial names by chemists?
    b) Look carefully at the systematic names of fatty acids and deduce what you can about the significance of each number to the right of the name.

3. **a)** Draw the skeletal formula of stearic acid.
    b) Calculate the M_r of stearic acid and oleic acid.
    c) Even though they have similar relative molecular masses, stearic acid is a fat and oleic acid is an oil. Oleic acid is a Z-isomer, also called a *cis*-isomer. It has the skeletal formula shown in Fig 63.

    Explain this difference in melting points by considering the action of the relevant intermolecular forces. (Hint: Look at the skeletal formulae you drew to answer Question 3a.)

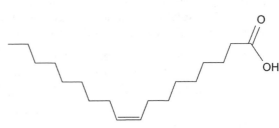

△ Fig 63 Z-oleic acid.

Oils can be changed to fats simply by removing some of the double bonds and this is how margarine was developed. In the 20th century, hydrogenated vegetable oils found their way into many processed foods. Other low fat spreads were also invented as scientific research revealed the link between saturated fat consumption and health risks.

A major drawback to hydrogenating vegetable oils is the production of *trans* fatty acids (*E*-fatty acids). Naturally occurring fatty acids are almost all *cis*-isomers (Z-fatty acids). In the 1980s scientific research first began to suggest that *trans* fats in the diet might lead to a much increased risk of heart disease. Further studies pointed to a causal link. It appears that *trans* fatty acids increase so-called "bad" cholesterol in our blood. This type of cholesterol coats the insides of blood vessels. It also decreases the body's production of "good" cholesterol, which scours blood vessels removing fatty deposits. Other studies have even suggested that *trans* fats promote obesity.

Once scientific evidence started to suggest a causal link between heart disease and *trans* fatty acids, governments began to insist that food labels should clearly state the proportion of *trans* fats they contained. This has led to food companies finding ways to make their products without partially hydrogenated fats. However, the causal link between *trans* fatty acids and significant health risks is still disputed.

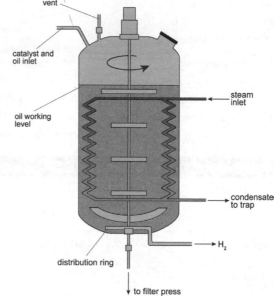

△ Fig 64 Apparatus used to hydrogenate vegetable oils.

QUESTIONS

4. **a)** Draw the skeletal formula of the *trans*-isomer (*E*-isomer) of oleic acid, showing clearly the arrangement around the carbon-carbon double bond. Use Fig 63 as a guide.

    b) Explain why *trans* fatty acids (*E*-fatty acids) cannot be metabolised in the same way as *cis*-isomers. (Hint: Consider the diagram you have drawn for 4a) and Fig 64.)

5. **a)** What catalyst is used in the hydrogenation of vegetable oils?
    b) Explain the mechanism by which this solid catalyst speeds up hydrogenation.

6. How could hydrogenation be used to find out the number of double bonds in a pure unknown fatty acid of known molecular mass?

7. When fats, such as butter, turn rancid, the double bonds in the fatty acids are broken in a complex oxidation reaction. This produces short-chain carboxylic acids, which are responsible for the unpleasant odour of rancid fats. Linoleic acid is one of these fatty acids. Draw the displayed formulae of two possible short-chain carboxylic acids that could be formed from the oxidative splitting of linoleic acid.

8. Write a newspaper article explaining how governments, manufacturers and the public have used the research about *trans* fatty acids and *trans* fats to inform decision making and to what extent this is justified. Your article should include an explanation of a causal link. **QWC**

Skills practised
8, 10, 12, 13, 14, 15, 18, 19, 25, 26, 27, 28, 31, 39

A15 ARAMIDS: FIRE-RESISTANT AND BULLETPROOF

Nylon-6,6 is a polyamide. When it was first invented it was found to have three very important properties. Its fibres have a high tensile strength, it has a melting point of about 263 °C and it softens at relatively high temperature. It quickly found uses in some machine parts. However, it can only be in continuous use at the much lower temperature of 120 °C.

△ Fig 65 Like all F1 racing drivers, Fernando Alonso wears a suit lined with Nomex®.

Researchers at the chemical company DuPont have been at the forefront of developing aramids. Aramids, as their name suggests, involve benzene rings (*arenes*) linked by *amide* functional groups. Their first commercial success was Nomex®. It has a melting point above 400 °C. Its main application is as a fire-resistant material. The suits of racing drivers and the tunics of firefighters are lined with it.

The tensile strength of Nomex® is an improvement on that of nylon. However, Kevlar®, another aramid developed by DuPont is much better. A 7 cm diameter steel cable has a breaking strain of about 40 tonnes, the same diameter cable made from Kevlar® fibres is just as strong but very much lighter. It can be used instead of steel to reinforce the tyres of heavy vehicles but another very important use is for the stuffing in bulletproof vests.

△ Fig 66 Spanish policeman wearing a Kevlar® bulletproof vest.

QUESTIONS

1. The repeat unit of nylon-6,6 is shown in Fig 67. It is formed by two different monomer molecules. One of them is a dicarboxylic acid.
   a) Draw the displayed formula of each monomer.
   b) The average M_r of nylon-6,6 is 1.13×10^5. Calculate the number of repeat units in the polymer.
   c) Why is the M_r of this polymer an average value?

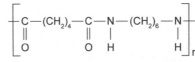

△ Fig 67 The repeat unit of nylon-6,6.

2. Nomex® is made from the two monomers shown in Fig 68:
   a) Name these functional groups: i) COCl; ii) NH_2
   b) Draw a section of the polymer chain showing two repeat units.
   c) What is the functional group of the polymer?
   d) What type of polymer is Nomex®?

3. Kevlar® can be made from 1,4-diaminobenzene and 1,4-benzenedicarboxylic acid. Draw the structures of these monomers and the repeat unit of Kevlar®.

4. Explain why Kevlar® is so much stronger and has a higher melting point than Nomex®. Your answer should explain the relevant intermolecular force that holds these polymer strands together. **QWC**

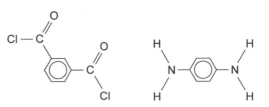

△ Fig 68 The two monomers of Nomex®.

Skills practised
25, 26, 27, 28, 29, 31, 32, 33, 39

A16 REACTION KINETICS AND VEHICLE EXHAUSTS

Even with the catalytic converter, vehicle exhaust pollution remains a major health issue, particularly in large towns and cities. Beijing, Los Angeles and Athens are infamous examples. On hot sunny days, with no wind, the build-up of NO_x (a mixture of NO and NO_2) plays an essential role in creating the pollutants present in photochemical smog. During these days there is a notable increase in the number of people dying.

△ Fig 69 Photochemical smog in Beijing in 2012, created largely from vehicle exhaust pollution.

Chemists can begin to suggest mechanisms for pollution reactions through studying their rates. Once rate equations are worked out, mechanisms can be proposed. Understanding the mechanisms of reactions can lead to better solutions to vehicle pollution.

QUESTIONS

1. The average half-life for the removal of NO_x from the atmosphere in large cities during daylight hours is 3.8 hours. The half-life is constant.

   a) What is the order of reaction with respect to NO_x?
   b) On a particular day, the concentration of NO_x in the atmosphere is 4.24×10^{-4} mol dm³. How many daylight hours will it take this to reduce to an estimated safe limit of 5.30×10^{-5} mol dm⁻³?

2. Nitrogen monoxide reacts with oxygen to form nitrogen dioxide:

   $$2NO(g) + O_2(g) \rightarrow 2NO_2(g)$$

   A chemist carries out a series of experiments and determines the rate equation for this reaction:

   $$\text{rate} = k[NO(g)]^2[O_2(g)]$$

   In one experiment the chemist reacted 2.00×10^{-3} mol dm⁻³ NO(g) with 3.00×10^{-3} mol dm⁻³ $O_2(g)$. The initial rate of reaction was 1.60×10^{-3} mol dm⁻³ s⁻¹.
   a) What is the overall order of this reaction?
   b) Calculate the rate constant, k, for this reaction, giving its units.

c) Predict the effect of the following concentration changes on the initial rate of reaction:
  i) Doubling the concentration of nitrogen monoxide.
  ii) Halving the concentration of oxygen.
  iii) Increasing the concentration of both reactants by four times.
d) This reaction is thought to take place by a two-step reaction mechanism, with the first step being slower. Suggest a two-step mechanism for the reaction.

3. The following reaction takes place in car exhaust gases:

$NO_2(g) + CO(g) \rightarrow NO(g) + CO_2(g)$

An experiment to determine the rate equation has the results shown in Table 10.

Experiment	$[NO_2(g)]/mol\ dm^{-3}$	$[CO(g)]/mol\ dm^{-3}$	Initial rate/mol $dm^{-3}s^{-1}$
1	4.80×10^{-1}	2.52×10^{-2}	6.40×10^{-6}
2	2.40×10^{-1}	2.52×10^{-2}	3.20×10^{-6}
3	4.80×10^{-1}	5.04×10^{-2}	1.28×10^{-5}

△ Table 10

a) Deduce the order of reaction with respect to each reactant at 500 K.
b) Write the rate equation.
c) Determine the overall order for the reaction.
d) Work out the value for the rate constant, showing its units.

4. At temperatures less than 500 K the rate equation for the reaction in Question 3 has been found by experiment to be:

rate = $k[NO_2]^2$

Two mechanisms have been proposed for this reaction:

Mechanism 1: Step 1 $NO_2 + NO_2 \rightarrow NO_3 + NO$ *slow*

  Step 2 $NO_3 + CO \rightarrow NO_2 + CO_2$ *fast*

Mechanism 2: Step 1 $NO_2 \rightarrow NO + O$ *slow*

  Step 2 $CO + O \rightarrow CO_2$ *fast*

Explain in detail which mechanism best agrees with the experimental evidence. **QWC**

Skills practised
8, 12, 16, 25, 26, 27, 28, 29, 30, 31, 32, 34

Assessing Investigative Skills

AIS1

One way to remove limescale from the heating element of an electric kettle is to use a dilute solution of sulfamic acid. One commercial product is a solution of this acid. It is tested by a student using volumetric analysis.

- The student uses a pipette and filler to transfer 25.0 cm³ of the sulfamic acid product into a 250 cm³ volumetric flask and adds distilled water up to the 250 cm³ mark.
- The same pipette is rinsed with the diluted solution of the sulfamic acid product.
- The pipette and filler are now used to transfer 25.0 cm³ of the diluted product into a conical flask.
- Thymolphthalein indicator is added to this solution and it remains colourless.
- A burette is filled with 0.100 mol dm⁻³ sodium hydroxide (NaOH(aq)).
- The endpoint is when one drop of sodium hydroxide solution turns the indicator pink.
- The burette readings are recorded to the nearest 0.05 cm³ in Table 11 below.

Titration number	1	2	3	4
Final burette reading/cm³	21.50	42.20	21.35	42.15
Initial burette reading/cm³	0.50	21.40	0.80	21.50

△ Table 11

Question 1

a) Draw this table and add a row to show the titres. (1)

b) Calculate the mean titre to the appropriate precision. (1)

The equation below represents the reaction that occurred. HA is sulfamic acid.

NaOH(aq) + HA(aq) → NaA(aq) + H₂O(l)

In the following calculations, show your working.

c) Calculate the amount, in moles, of NaOH used in the titration. (1)

d) Determine the amount, in moles, of sulfamic acid (HA) present in 25.0 cm³ of the diluted product solution. (1)

e) Calculate the amount, in moles, of sulfamic acid present in the 250 cm³ solution. (1)

f) Calculate the concentration, in mol dm⁻³, of sulfamic acid present in the original commercial product. Give your answer to three significant figures. (2)

g) What is:

    i) the maximum error in a single burette reading? (1)

    ii) the percentage error in one of the titre values in Table 11? (1)

h) When the pipette is used for the second time it is rinsed with the diluted solution the student had made.

   i) Explain why this is done. **(1)**

   ii) What would be the effect on the titre of not rinsing the pipette? **(1)**

i) Each time the student performs a titration she rinses out the conical flask with distilled water. What effect does this procedure have on the titre? **(1)**

j) An air bubble is present in the tip of the burette at the start of the first titration but not at the end. Explain why this makes the titre less reliable. **(1)**

Question 2

The student decides to determine the relative molecular mass of sulfamic acid. She uses the apparatus in Fig 70 to measure the volume of carbon dioxide produced when sulfamic acid reacts with 0.50 mol dm^{-3} sodium carbonate (Na_2CO_3) solution.

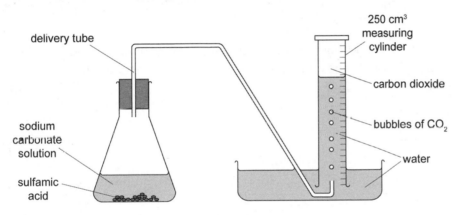

△ Fig 70 The apparatus used to measure the volume of carbon dioxide produced from the student's reaction.

The results she obtains are shown in Table 12.

Mass of conical flask and sulfamic acid	134.57 g
Mass of conical flask empty	132.87 g
Volume of sodium carbonate solution added	100 cm^3
Volume of carbon dioxide collected	207 cm^3

△ Table 12

a) What is the mass of sulfamic acid used? **(1)**

b) Using HA to represent sulfamic acid, write the balanced equation, including state symbols, for its reaction with sodium carbonate. **(2)**

c) How many moles of carbon dioxide are released? **(1)**

d) All the sulfamic acid reacts. How many moles of sulfamic acid are present in the weighed sample? **(1)**

e) Calculate the molar mass of sulfamic acid to 3 significant figures. (1)

f) The student evaluates the measurement of carbon dioxide. The volume has a maximum error of ± 1 cm^3.

Calculate the percentage error in the volume of CO_2 measured. (1)

g) The accurate value for the molar mass of sulfamic acid is 97.1 g. Give one procedural error that could result in the molar mass being higher than the accurate value. (1)

h) How could the student ensure that her results are more reliable? (1)

Question 3

a) Identify one hazard in the titration experiment and state what could be done to minimise the risk. (2)

b) Concentrated solutions of sulfamic acid are a hazard to aquatic life. How should they be disposed of to minimise the risk to the environment? (1)

Question 4

Sulfamic acid is composed of 3.09% hydrogen, 14.42% nitrogen, 33.06% sulfur and the remaining element is oxygen. Calculate the molecular formula of sulfamic acid using the accurate value for M_r in 2(g). (2)

(A_r: H = 1.0; N = 14.0; S = 32.1; O = 16.0)

Skills practised

3, 4, 5, 6, 7, 8, 9, 10, 11, 14, 16, 18, 19, 20, 23, 24, 25, 26, 27, 28, 29, 30, 31, 32, 33

Section A

Question 1

Magnesium reacts with dilute hydrochloric acid to give hydrogen gas.

$Mg(s) + 2HCl(aq) \rightarrow MgCl_2(aq) + H_2(g)$

A student wishes to determine the order of this reaction using the apparatus in Fig 71.

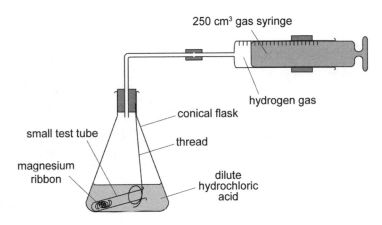

250 cm³ gas syringe

hydrogen gas

conical flask

small test tube

thread

magnesium
ribbon

dilute
hydrochloric
acid

△ Fig 71

a) The student predicts that the reaction is second order with respect to HCl(aq).

rate = $k[HCl(aq)]^2$

Sketch a graph that displays this prediction. Label the axes and include units. **(2)**

b) The student decides that the maximum volume of gas he will collect is 240 cm³ at 298 K and 101 kPa. The concentration of hydrochloric acid he has been given is 0.50 mol dm⁻³.

i) What volume of this acid should he use? **(3)**

ii) Why must the magnesium be in excess? **(1)**

c) The student plans to do five experiments to test his prediction, using five different concentrations of HCl(aq). He constructs a table (see Table 13).

Expt. no.	Volume of 0.50 mol dm⁻³ HCl(aq)	Volume of H₂O	Concentration of HCl(aq)
1	40	0	0.50
2			
3			
4			
5			

△ Table 13

i) Copy out Table 13 and complete it to show how you would prepare four more solutions of hydrochloric acid. Make sure that you add the correct units. **(2)**

ii) Identify the independent variable and the dependent variable in the experiment the student plans. **(2)**

iii) Using the concentrations you have already calculated in Table 13, write your own plan for this experiment, with a step-by-step description of each stage. **(3)**

d) The student calculates the initial rate of reaction for each of the five experiments to test if his prediction is correct. Table 14 shows one set of results the student obtains from the experiment using 0.50 mol dm^{-3} HCl(aq).

i) Plot a graph of these values. **(1)**

ii) Calculate the initial rate of reaction for this experiment. **(1)**

Time/s	Volume of H^2/cm^3
0	0
20	48
40	88
60	128
80	152
100	176
120	192
140	208
160	216

△ Table 14

Section B

Question 2

The sugar that may be used to sweeten tea or to cook with is called sucrose. It is a disaccharide made up of the monosaccharides glucose and fructose, which both have the same molecular formula $C_6H_{12}O_6$. In acidic solution sucrose is hydrolysed.

$$C_{12}H_{22}O_{11}(aq) + H_2O(l) \rightarrow C_6H_{12}O_6(aq) + C_6H_{12}O_6(aq)$$

sucrose   glucose   fructose

A chemist follows this reaction to determine its order. Samples of the sucrose solution are analysed at regular intervals during the course of the reaction and the concentrations of sucrose recorded.

The chemist predicts that this reaction will be first order with respect to sucrose and he uses this equation to check his prediction.

$$\log_{10} [\text{sucrose(aq)}] = -kt + \log_{10} [\text{initial sucrose(aq)}]$$

k is a constant and t is the time.

If it is a first order reaction then a plot of $\log_{10}$ [sucrose(aq)] against time will produce a linear graph.

Results from the chemist's experiment are shown in Table 15.

a) Draw this table and add a third column. (2)

Process the results to enable you plot a graph of $\log_{10}$[sucrose(aq)] against time t. You should label the third column appropriately and include any units. You should give the values to 3 significant figures. **(2)**

b) Plot a graph of $\log_{10}$ [sucrose(aq)] against time t and draw a line of best fit.

Remember that $\log_{10}$ [sucrose(aq)] values are negative. **(4)**

time/min	[sucrose(aq)]/mol dm⁻³
0	0.500
40	0.427
80	0.363
120	0.309
160	0.288
200	0.224
240	0.191
280	0.162

△ Table 15

c) On the graph:

i) Identify with a circle any points that are anomalous and give an explanation as to why any anomaly may have occurred. **(2)**

ii) How reliable is the data obtained for this experiment? Give your reasons. **(1)**

d) Use your graph to estimate the concentration of sucrose after 100 minutes. **(1)**

e) Graphs with linear relationships are represented by this equation:

$y = mx + c$

i) Identify y, m, x and c from the equation the chemist used to verify his prediction. **(1)**

ii) Calculate the slope of this graph showing clearly the construction lines you have used. Give your answer to three significant figures and include the units. **(3)**

f) Is the chemist's prediction that this is a first order reaction correct? Explain your answer. **(1)**

g) In performing this investigation the chemist must keep the temperature constant.

i) Can you explain why? **(1)**

ii) On your graph, draw another line that shows the effect of doing this experiment at a lower temperature. **(2)**

Skills practised

3, 4, 5, 7, 8, 9, 10, 11, 12, 18, 19, 20, 21, 22, 23, 25, 26, 27, 28, 31, 32, 34, 35, 36, 37, 38

Answers

Each activity's set of questions gives you the opportunity to practise and develop skills covered in the **Skills** section and, by answering them, you will enhance your understanding of the skills necessary for success, making your chemistry course easier to understand, your revision a more active process and exam questions much less daunting.

This section includes all the answers, as well as helpful hints and tips to boost your performance in that exam room. Many of the answers here show you exactly how to structure your responses and we have paid particular attention to the steps in calculations to make it as clear as possible what you need to do.

QWC The last question in each activity is a Quality of Written Communication (QWC) question. These test your subject knowledge and understanding and ask you to think about how you communicate your ideas. For these questions, we have not only outlined the points you need to cover but have also included an indication of low, medium and high scoring responses to show you how to improve. For more guidance on QWC questions, we have written low, medium and high scoring responses to A3, Question 4 in the **QWC Worked Examples** section. Each response has detailed points commenting on its quality of written communication, giving you an insight into what's needed.

The **Assessing Investigative Skills** section gives you the chance to develop your investigative skills in two major activities without the need to complete practical work. You will be able to tackle AIS1 towards the start of your course and AIS2 will help you tackle more advanced investigations later on. The answers to these offer comprehensive guidance so you can check your responses and see how to progress.

Lastly, remember that learning facts is only a small part of studying science – examiners and future employers are looking for someone who can analyse, think logically and apply their knowledge to new situations; practising the skills will develop your ability to do this.

For more hints on how to improve the quality of your written communication, see the QWC Worked Examples.

A1 Iron-60 found on Earth

1. 1 nm is 10^{-9} metres. 80 nm is 8×10^{-8} m in standard form.

2. The easiest way to answer this is to imagine that you have 100 g of iron oxide (see Table 16).

	Iron	Oxygen
Mass in grams	69.9	30.1
Amount in moles	$\frac{69.9}{55.9} = 1.25$	$\frac{30.1}{16.0} = 1.88$
Simplest ratio (Divide by the smallest number)	$\frac{1.25}{1.25} = 1.00$	$\frac{1.88}{1.25} = 1.50$
Simplest whole number ratio	2	3

△ Table 16

So the empirical formula is Fe_2O_3. $(55.9 \times 2) + (16.0 \times 2)$ = 159.8, which is the M_r. Formula is Fe_2O_3.

3. A_r (iron) =

$$\frac{(54 \times 5.84) + (56 \times 91.68) + (57 \times 2.17) + (58 \times 0.31)}{100}$$

$$= \frac{5591.11}{100} = 55.91$$

4. a) Before a letter or paper is published it is usually subject to peer review and so will be sent to other scientists with an expertise in the field. These scientists check to see if the results are reliable and valid and if the conclusions are valid, before publication can occur.

   b) The Activity describes *traces* of iron-60 so there will be some doubt in the minds of other scientists that the data is reliable and valid. A larger sample means the experiment can be repeated with more precision to see if similar results are obtained, thus confirming their reliability.

5. **QWC** You are asked to explain how this activity describes scientists working scientifically. When you plan your answer, try to jot down the significant points from the description you are given. Remember that this is a QWC question so punctuation and grammar must be accurate and the answer well organised. There are 10 points that could be made and you may think of others. A low level answer will only include two or three of these points and they will probably not be clearly written. A medium level answer will cover four or five points with more clarity. A high level answer will have covered at least six points in a clear and well-structured answer.

   - A scientist observes evidence for a mass extinction two million years ago.
   - The scientist puts forward a hypothesis/idea that a supernova could have caused this.
   - Since we know about the hypothesis, he must have communicated it in a journal.

- At this point there is no data to support the hypothesis.
- Other scientists put forward their hypothesis that radioactive atoms in fossils could provide evidence for supernovae explosions.
- A radioactive isotope is found that should not be present on Earth.
- This could confirm the original hypothesis but the amount is very tiny.
- The results are published in a scientific journal.
- Another scientist reads this and suggests where the German team might look for more reliable data.
- When scientists cooperate together science advances more quickly.

A2 The nuclear model of the atom

1. The hazard is from radiation emitted from radio-active materials. Nowadays experiments using radioactive materials are strictly controlled, but in 1912 the dangers were not well understood. Here are some points – any one of them would score the mark. There are many more.
    - Avoid contamination of personal clothing and skin by wearing protective clothing that can be safely destroyed.
    - Shower after doing experiments with radioactive material.
    - Use shields around the apparatus to absorb radiation.
    - Keep exposure to radioactive material to a minimum by planning experiments carefully.
2. The alpha particles that were reflected back.
3. To ensure the data was reliable and that the same anomalous results were repeated.
4. **a)** Most of the alpha particles passed through undeflected.
    b) Only a very few alpha particles were deflected. This suggests
    - a very small volume for the nucleus
    - gold atoms are much heavier than alpha particles so most of this mass must be in this very small volume.
    c) The alpha particles that were reflected back. Alpha particles are positively charged so they must have been repelled by a positive charge on the nucleus.
5. **QWC** This question has all of the information in the activity. You first need to extract the relevant points then organise them well. You also have to show how each point relates to science advancing.

Do not miss out the other part of the question about explaining observations scientifically. These are shown in the points with an asterisk.

A high level response will answer both parts of the question so it will include at least three of the points with an asterisk together with four or more of the other points. It will also use some, or all, of the underlined scientific terms correctly.

There are a lot of points here:

- A scientist proposes a <u>model</u> to explain observations.
- As long as the model explains all the known observations then it is accepted by the scientific community.
- Thomson proposed his model because he had just discovered the first subatomic particle. He believed that the electron was the only sub-atomic particle.
  * Thomson knew that atoms were neutral so the negative charge he observed on the electron had to be balanced by a sphere of positive charge. This was part of his scientific explanation.
  * The electrons are very light so thousands were needed to explain the masses of atoms. This was another part of his scientific explanation.
- When Rutherford fired alpha particles at gold atoms his <u>prediction</u> was that Thomson's model should have allowed them all through with very little deflection.
- The <u>anomalous results</u> that were observed were repeated many times so Rutherford believed they were *reliable*.
- Rutherford rejected Thomson's model because it could not explain the reflected alpha particles.
  * Rutherford proposed a new model to explain the observations scientifically.
  * The nucleus is positive because it repels positive alpha particles.
  * The nucleus is tiny and has almost all the atom's mass because most alpha particles are undeflected.
  * The positive charge is balanced by negative electrons.

A3 The arrangement of electrons in shells

1. $Mg^{11+}(g) \rightarrow Mg^{12+}(g) + e^-$
    The state symbols are essential here.

Number of electron removed/mol	1	2	3	4	5	6	7	8	9	10	11
Successive IE/kJ mol^{-1}	500	4600	6900	9500	13 400	16 600	20 100	25 500	28 900	141 000	159 000
log$_{10}$ of successive IE	2.70	3.66	3.84	3.98	4.13	4.22	4.30	4.41	4.46	5.15	5.20

Table 17 Answer to Q2a).

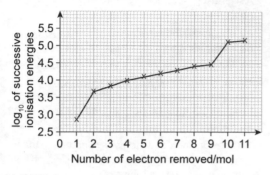

△ Fig 72 Answer to Q 2a).

2. **a)** See Table 17 and Fig 72. In plotting the graph in Fig 72 each point is joined by a separate straight line because the data is discrete.

b) The values range over several orders of magnitude, so using logarithms condenses the scale, making the numbers more manageable and patterns easier to see.

c) The trend is of increasing ionisation energies as each successive electron is removed.

- There is a large jump in energy required to remove the second electron. This is because we are breaking into a new shell, which lies closer to the nucleus.

- Also the first electron removed has more inner shells shielding it from the positive nuclear charge.

- The next eight electrons are removed from the second shell.

- The 10th electron is removed from the first shell, which is closest to the nucleus, and there is another large jump in ionisation energy.

3. The answer to this question is in the QWC Worked Examples section.

A4 The arrangement of electrons in subshells

1. **a)** $Li(g) \rightarrow Li^+(g) + e^-$

b) 520.2 kJ mol^{-1} is the energy to remove 1 mole of electrons from 1 mole of Li atoms.

There are 6.02×10^{23} electrons in 1 mole.

The energy required to remove 1 electron

$$= \frac{520.2}{6.02 \times 10^{23}} = 8.64 \times 10^{-22} \text{ kJ to 3 sf.}$$

Three significant figures are given because the value of the mole is to 3 sf.

2. The general trend in ionisation energies is that they increase across the period.

- Across a period the nuclear charge increases because the number of protons increases.

- This means the electrons are more strongly attracted across the period.

- Therefore the energy required to remove the first electron increases.

3. **a)** See Fig 74.

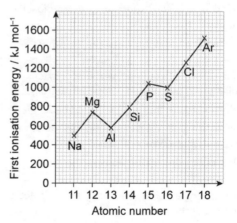

△ Fig 73 First ionisation energies of successive elements in Period 3.

b) Both graphs show a general trend of increasing ionisation energies with a 2,3,3 pattern.

4. **QWC** You have been asked to explain various parts of the graph. A high level answer will have all the points detailed for each section. A medium level response will probably not include the asterisked points. A low level response may miss out the points in c) altogether.

a) There is an increase in ionisation energy because Be has an extra proton.

* This attracts beryllium's electrons more strongly.

b) Boron shows a decrease in ionisation energy because it has three electrons and the one that is removed is in the 2p subshell.

* The 2p subshell is at a higher energy level than the 2s subshell.

c) Oxygen has four electrons in the 2p subshell.

* This is the first electron to be paired up so they repel each other, making this electron easier to remove.

A5 Noble gases react!

1. First ionisation energies increase across a period and reach a maximum at Group 0, the noble gases. Removing an electron from this stable electron configuration requires the most energy of any element in a period.

2. See Fig 74

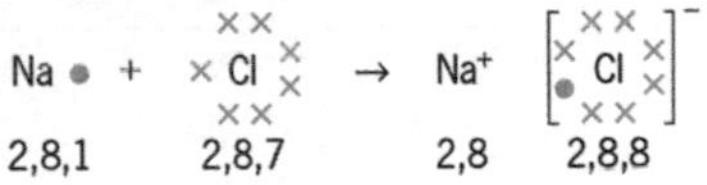

$$Na \bullet \; + \; \times \overset{\times\times}{\underset{\times\times}{Cl}} \times \; \rightarrow \; Na^{+} \; \left[\overset{\times\times}{\underset{\times\times}{Cl}} \times \right]^{-}$$

2,8,1   2,8,7   2,8   2,8,8

△ Fig 74 Dot and cross diagrams for the formation of sodium and chloride ions.

3. See Figs 75, 76 and 77.

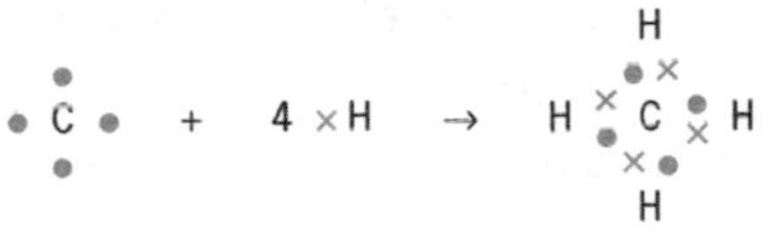

$$\bullet \overset{\bullet}{\underset{\bullet}{C}} \bullet \; + \; 4 \times H \; \rightarrow \; H \overset{H}{\underset{H}{\times C \times}} H$$

△ Fig 75 Dot and cross diagram for methane.

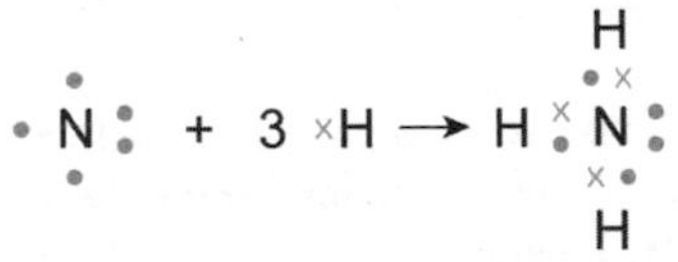

$$\bullet \overset{\bullet}{N} \; + \; 3 \times H \; \rightarrow \; H \overset{H}{\underset{H}{\times N \bullet}}$$

△ Fig 76 Dot and cross diagram for ammonia.

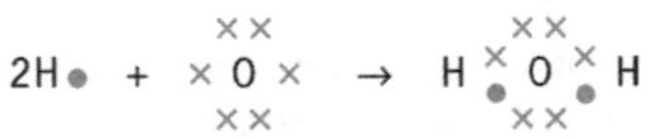

$$2H \bullet \; + \; \times \overset{\times\times}{\underset{\times\times}{O}} \times \; \rightarrow \; H \overset{\times\times}{\underset{\times\times}{O}} H$$

△ Fig 77 Dot and cross diagram for water.

4. In BF_3, boron has only six electrons in its outer shell and this is two less than a noble gas outer shell of eight. See Fig 78.

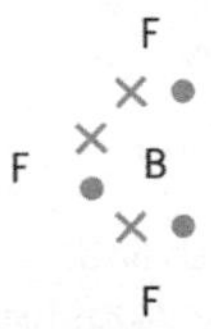

$$F \overset{F}{\underset{F}{\times B}}$$

△ Fig 78 Dot and cross diagram for boron trifluoride.

The SF_6 molecule has a sulfur atom surrounded by 12 outer shell electrons. This is four more than a noble gas octet. See Fig 79.

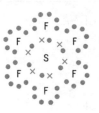

△ Fig 79 Dot and cross diagram for sulfur hexaflouride.

5. a) [QWC] In this section a high level response will include at least two of the bullet points.

A medium level response may include two points but there will be less clarity in the answer. A low level response may make one of these points but the underlined words will probably be missing.

* Noble gases were believed to be totally unreactive because of their stable outer shells and this theory was accepted by the scientific community.
* *Nature* is a well-respected scientific journal and the editor would probably have been sceptical.
* The editor of *Nature* would probably have wanted the findings peer-reviewed.

b) In a situation like this, a moment spent thinking through scientific working will enable you to make some sensible points. A high level answer will make two of these. A medium level answer may well make two but the organisation of the answer could detract from the overall mark. A low level answer may miss all of these points.

* It would be difficult to keep such a discovery secret for long.
* Neil Bartlett would have wanted the credit for discovering that noble gases could react to form a compound.
* Other chemists may have realised that a very powerful oxidising agent might make a noble gas react.
* Neil Bartlett would want other scientists to reproduce his results.

A6 The search for new medicines

1. The *lock and key* model of drug action. This model is useful because the idea of a key fitting into a lock and turning helps us to understand in a simple way how drugs act.

2. Tables are a very good way of presenting data. For this table (see Table 18) you are being asked to organise information in a way that makes it easy to

Compound	Number of pairs of electrons	Dot and cross diagram	Drawing of the molecule to show bond angles	Name of shape
$BeCl_2$	2	Cl Be Cl (dot and cross diagram)	Cl —— Be —— Cl, 180°	linear
BCl_3	3	Cl, B, Cl, Cl (dot and cross diagram)	Cl with B centre, 120° 120° 120°	trigonal planar
CH_4	4	H C H with H top and H bottom (dot and cross diagram)	H, C, H, H with 109.5° 109.5°	tetrahedral

△ Table 18 Answer to Q 2. The shapes of some molecules.

understand. Before you start, plan out how many columns and rows you are going to need and how you will label them.

3. Four bonds are visible around each of the blue atoms. This means four bond pairs, so the other atoms will be tetrahedrally arranged and the bond angle will be 109.5°.

4. See Fig 80. With four bond pairs an angle of 109.5° is expected. Since 107° is less than this, the lone pair must have a greater repulsion than the bond pairs.

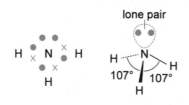

△ Fig 80 The ammonia molecule.

lone pair

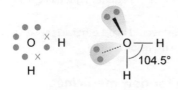

△ Fig 81 The water molecule.

Again (see Fig 81) there are four bond pairs but the bond angle is even less than in ammonia. The two lone pairs must repel each other more than the bond pairs.

5. **QWC** When considering the points you are going to use, always look for clues in the information you are given. Here are two:

- *easily visualised using computer graphics*
- *previously this was all done in the laboratory.*

This tells you that time is a factor.

Reasons for not using a lab:

- Test-tube reactions are very time consuming and therefore very expensive.
- Only a very few drug molecules can be tested at any one time.
  * The shape of the drug molecule and active site are much more difficult to interpret in two dimensions on paper.
  * Many potential drug molecules will not react with the active site, wasting a lot of time.

Reasons for using computer graphics:

  * The shapes of a drug molecule and active site can be seen in three dimensions.
- It is much easier and quicker to see if potential drug molecules will actually fit a receptor site.
- Because it is quicker, many more potential drug molecules can be tested in a day.
- The speed of testing the drug molecules makes it much cheaper.
  * Only drugs that look as if they may fit the receptor site need to be tested.

You may be able to think of other points.

A high level answer will *compare* the lab and computer graphics for testing possible drugs using similar points from each list. Two good comparisons and an additional point should score highly. The asterisked points also indicate a higher level answer but a well-organised response may still score well.

A medium level answer may not give clear comparisons and will probably not mention the asterisked points.

A low level answer will be disorganised and there may only be one comparison.

A7 Bioethanol and Brazil

1. *Biofuels* are fuels made from vegetable matter.

   Carbon-neutral fuels are those that only give out the same carbon dioxide when burnt as was taken in by the plants to make the carbon compounds that are used for the fuel. There is no net increase in carbon dioxide in the atmosphere.

2. **a)** All the oil that is burnt as fuel releases additional carbon dioxide into the atmosphere from the carbon compounds that are already in it. This carbon dioxide was removed from the atmosphere millions of years ago.

   It requires energy and therefore adds more carbon dioxide to the atmosphere to extract, refine and transport fuels from oil. If this comes from fossil fuels then it is yet more carbon dioxide released into the atmosphere.

   Growing sugar cane means that some carbon dioxide from the atmosphere is removed by the cane to make the sugar but the juice still needs to be extracted, processed and transported, releasing some carbon dioxide saved back into the atmosphere.

   b) Whether the sugar is produced for food or fuel it will take energy to plant it, grow it (if fertilisers are used), harvest it and process it. All these stages produce carbon dioxide.

   Additional energy and therefore carbon dioxide will be released into the atmosphere to replace the crops that are now not being produced for food.

3. **a)** Mass of 1 mol C_2H_5OH =

   $(12 \times 2) + (1 \times 5) + 16 + 1 = 46$ g

   Moles of ethanol in 1kg = $\dfrac{\text{mass in grams}}{\text{mass of one mole (in grams)}}$

   $= \dfrac{1000 \text{ g}}{46 \text{ g}} = 21.739$ mol

   1 mole ethanol produces 1370 kJ of energy

   21.739 mol ethanol produces 1370×21.739
   $= 29\,782$ kJ

   Energy density of ethanol = 29 800 kJ kg^{-1} to 3 sf

   b) i) $29\,800 \times \dfrac{155}{100} = 46200$ kJ kg^{-1}

ii) Number of litres of petrol in 1 kg

   $= \dfrac{1000 \text{ g}}{740 \text{ g}} = 1.35$ litres

iii) For every 1 litre of petrol you would need to purchase 1.55 litres of ethanol to get the same energy when the fuel is burnt.

   Percentage $= \dfrac{1}{1.55} \times 100 = 64.5\%$, so ethanol would need to be at least 35.5% cheaper per litre.

4. **QWC** A well-organised response that includes two advantages and two disadvantages would be a high level answer.

   A medium level answer may be less well organised and contain three points.

   A low level answer may not cover both advantages and disadvantages, or some of the points will be irrelevant or incorrect.

   Advantages

   • Biofuels are renewable/sustainable, or the reverse argument.

   • Oil that would be burnt is used to produce important chemicals.

   • Large-scale pollution from oil spills is avoided.

   Disadvantages

   • Food grown for fuel is not eaten, so more land has to be found for food crops OR there will be a shortage of land for growing food.

   • Forests may be cut down to make more land available. *Note:* A comment that just mentions loss of habitat will not be enough.

   • The most fertile land may be used for fuel crops OR the fertility of the land used for fuel crops may be damaged.

A8 Hydrogen: car fuel of the future?

1. Mass of 1 mole MgH_2 = 24.3 + (1.0 × 2) = 26.3

   Percentage of hydrogen in MgH2

   $= \dfrac{2.0}{26.3} = 7.6\%$ to 2 sf

2. **a)** $MgH_2(s) + 2H_2O(l) \rightarrow Mg(OH)_2(s) + 2H_2(g)$

   b) Step 1 Convert the equation to amounts: 1 mol MgH_2 gives 2 mol H_2

   Step 2 Work out the amount being used:

   Moles of MgH_2 in 1 kg $= \dfrac{1000 \text{ g}}{26.3} = 38.02$ mol (ignoring sf)

   Step 3 Scale the amounts in the equation:

   38.02 mol MgH_2 produces 2 × 38.02 mol H_2
   $= 76.04$ mol H_2

   Step 4 Calculate the volume of gas:

   1 mole of any gas occupies 24 dm^3 at 298 K and 101 kPa.

Volume of H_2 = 76.04 × 24 = 1825 dm³ per kilogram.

3. Van der Waals (induced dipole-induced dipole) forces because the carbon nanotube molecule is non-polar and so is hydrogen.

4. • The equilibrium can be written as:
H_2(adsorbed) $\rightleftharpoons$ H_2(g).

   • As the pressure falls, <u>equilibrium shifts to the right</u> to increase the pressure again.

   • This <u>minimises</u> the pressure decrease.

   • This produces more hydrogen.

For the answer to score full marks all the points above will need to be made. The underlined words should be present.

5. **QWC** You are asked to give advantages and disadvantages so your answer should include a balance of points from both these lists.

   Advantages
   * No harmful products produced during combustion.
   * No carbon dioxide produced so no contribution to global warming.
   * It has a very high energy density OR it gives out a lot of energy per gram. (In fact per gram it releases more energy on combustion than petrol.)
   • There is an almost unlimited supply of hydrogen in water.

   Disadvantages
   * The initial manufacture of hydrogen requires energy. If this comes from fossil fuel then this still produces carbon dioxide and uses up non-renewable fossil fuel.
   • Transporting and storing hydrogen give major safety concerns.
   * Storing hydrogen as a liquid requires energy that may come from fossil fuel.
   • The cost of changing the existing infrastructure would be incredibly high.
   • The political and public will to change.
   • Hydrogen is seen by the public as highly flammable and explosive. The same applies to petrol but there is a negative image that would have to be changed.

   These are some of the major points. You would be expected to give two from each list for a high level answer, including at least one asterisked point from each. The answer needs to be well structured, with correct spelling and punctuation.

   A medium level response still requires points from both lists and at least one asterisked point.

   A low level answer may not include any of the starred points, will not be well organised and probably will have major errors of grammar and punctuation.

A9 Fullerenes – new forms of carbon

1. a) A_r of carbon is 12.0. 12 × 60 = 720

   b) A positive ion, C_{60}^+.

      The C_{60} gas is bombarded by high-energy electrons to ionise the molecules by knocking electrons from them.

      This could also be shown by an equation:

      $$C_{60}(g) + \underset{\substack{\text{high-energy}\\\text{electron}}}{e^-} \rightarrow C_{60}^+(g) + e^- + e^-$$

   c) $\frac{840}{12}$ = 70. So C_{70}^+. The positive charge is essential as the question asks for 'species'.

   d) C_{60} produced a large peak so it did not break down in the mass spectrometer.

2. a) Carbon reacts with oxygen at high temperatures to give carbon dioxide.

   b) See Figs 82 and 83.

Every carbon atom bonds to four other carbon atoms

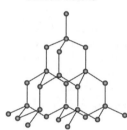

△ Fig 82 The structure of diamond.

Flat sheets of carbon atoms are bonded into a hexagonal structure

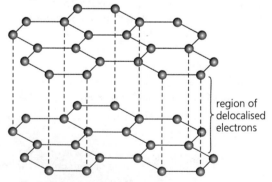

region of delocalised electrons

△ Fig 83 The structure of graphite.

For graphite, we have drawn six hexagons in each layer. A total of just two in each layer would be quite acceptable when answering this question.

The graphite structure consists of hexagons of carbon atoms and there are similar hexagons in fullerene.

3. a) A milligram is one thousandth of a gram, 0.001 g.
To convert milligrams to grams divide by 1000.

$$\frac{0.0306}{1000} = 0.000\,030\,6 = 3.06 \times 10^{-5}\,\text{g}$$

b) Step 1 Write the balanced equation:

$60C(\text{graphite}) \rightarrow C_{60}$

Step 2 Convert the equation to amounts:

60 moles C gives 1 mole C_{60}

Step 3 Work out the amount being used:

$$\text{Moles C(graphite)} = \frac{\text{mass in grams}}{\text{mass of one mole in grams}}$$

$$= \frac{1.208}{12.0} = 0.100\,67 \text{ ignoring sf}$$

Step 4 Scale the amounts in the equation:

$0.100\,67$ mol C gives $\dfrac{0.100\,67}{60}$ mol C_{60}
$= 0.001\,678$

Step 5 Convert amount of C_{60} to mass:

$0.001\,678$ mol $\times 720 = 1.208$ g

$$\text{Percentage yield} = \frac{\text{actual yield}}{\text{theoretical yield}} \times 100$$

$$= \frac{3.06 \times 10^{-5}}{1.208} \times 100 = 2.53 \times 10^{-3}\%$$

4. **QWC** All the points you need to make are in the Activity, so planning and organising your response is essential to a high level answer.

Remember that the audience for your response is the examiner and you should use appropriate scientific terms in the correct context for a high level answer. These are underlined in the points below.

- The amount of C_{60} produced in Kroto's original experiment was extremely small so the <u>reliability</u> of the data is suspect. Kratschmer produced a larger amount of C_{60} so results he obtained were more <u>reliable</u> and <u>valid</u>.
- Kroto's experiment relied on specialised equipment only available to the American scientists so the results were not <u>reproducible</u>. Kratschmer's experiment allowed other scientists to reproduce the results.
- Kratschmer's experiment produced enough C_{60} to allow analysis of a sample to determine its structure.
- Kroto's <u>hypothesis</u> about the structure of C_{60} was confirmed, meaning that his conclusion was valid.

You may think of other points. Three sensible points using three of the underlined words correctly would score very highly.

A medium level answer would cover two of the points but may only use one underlined word correctly.

A low level answer is likely to be disorganised and will give at least one of the points without clearly using the underlined words.

A10 Manufacturing nitric acid – a greener way

1. Mass of one mole $NH_4NO_3 = 80.0$ g. Mass of nitrogen in one mole $= 14.0 \times 2 = 28.0$ g.

$$\text{Percentage composition of nitrogen} = \frac{28.0}{80.0}$$

$$= 35.0\%$$

2. Worked answer

Step 1 Write the balanced equation, even if it is in the question.

$4NH_3 + 5O_2 \rightleftharpoons 4NO + 6H_2O$

Step 2 Convert the equation to amounts. In this case you are only interested in the moles of ammonia and the moles of nitrogen dioxide.

4 mol $NH_3 \quad \rightarrow 4$ mol NO

so 1 mol $NH_3 \quad \rightarrow 1$ mol NO

Step 3 Work out the amount that should be produced. This is called the *theoretical yield*.

$320\,000$ mol $\quad \rightarrow 320\,000$ mol

Step 4 Work out the amount that is actually produced. This is the *actual yield*.

$$\text{Moles of NO} = \frac{\text{mass in grams}}{\text{mass of one mole (in grams)}}$$

$$= \frac{9\,240\,000\,\text{g}}{30.0\,\text{g}} = 308\,000 \text{ moles}$$

Step 5 Calculate the percentage yield.

$$\text{Percentage yield} = \frac{\text{actual yield}}{\text{theoretical yield}} \times 100\%$$

$$= \frac{308\,000}{320\,000} \times 100 = 96.25\% = 96.3\%$$

The appropriate number of significant figures is 3 because the mass of nitrogen monoxide is quoted to 3 sf in the question.

3. When a system is in dynamic equilibrium, the <u>forward and reverse reactions are occurring at the same rate</u>.

The underlined part is the idea that must be there. A low grade answer will often miss out *the same rate* or forget to mention that it is the *rate* in *both directions* that is the same.

4. a)
- A manufacturer does not wait because time in industry is money.

- A lower yield can often be produced in a shorter time, so the amount produced per day is far more than waiting for equilibrium to be established.

- Unreacted gases are recycled to go through the reactor again.

- Both conditions shift the equilibrium to the left.

- The reaction rate is increased by high temperature so the products are obtained faster.

- A pressure of 1000 kPa is not high, but it means that the plant can have smaller pipes and reactors because there are more moles of reactants in a smaller volume.

- A higher pressure also increases the rate.

b) See Fig 84.

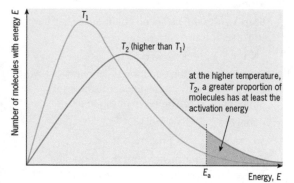

△ Fig 84 Answer to Q 4b). Boltzmann distribution curves at temperature T$_1$ and T$_2$.

Points to note when drawing your Boltzmann distributions:

- The axes require labels.

- The curves must start at the origin.

- They must not touch the *x*-axis on the right-hand side.

- The higher temperature curve is lower and to the right of the lower temperature curve.

- The curves only cross each other once and they should still be separate on the right-hand side.

You need to include:

- Higher temperature gives a faster rate of reaction.

- The higher temperature has more molecules above the activation energy *E*a.

- More successful collisions between molecules.

5. a) To increase the surface area of the catalyst and speed up the reaction rate.

b) None.

c) See Fig 85.

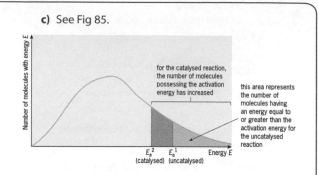

△ Fig 85 Answer to Q 5c). Boltzmann distribution curve showing the effect of adding a catalyst on the activation energy.

The relevant points about drawing a Boltzmann distribution are given in Question 4b).

Points to include:

- Using a catalyst increases the rate of reaction.

- The catalyst lowers the activation energy.

- More molecules have energy above activation energy.

6. a) The molecules absorb infrared radiation and this causes their bonds to vibrate. The molecules then release this energy, warming the atmosphere and the Earth's surface.

b) Any two from: infrared spectroscopy; mass spectrometry; gas chromatography; NMR.

c) You can usually tell how much to include in your answer by the number of marks allocated. Some points to include:

- Any change to existing plant has a capital cost, so legislation could tax the nitrous oxide emissions making it worthwhile to research into ways of modifying existing plant to reduce pollution levels.

- A new plant could be refused planning permission unless it complies with stricter emission limits.

- Public reputation is important to companies and if they are found to be breaking the law then this could damage sales. Their ethics will also be called into question.

7. The key aspect of this question is *international cooperation*, so your answer must address this in all the points you make. Some of these are:

- Nitrous oxide travels across national boundaries.

- One country on its own cannot have a large impact on reducing nitrous oxide levels.

- Scientists are encouraged to work together and share research findings.

- It allows the monitoring of nitrous oxide levels in different countries.

- Scientists can cooperate to warn governments about the risk.

- Global legislation may be produced.
- Richer countries are encouraged to assist poorer countries.

8. **QWC** A longer answer is required here and it is often worth jotting a few points down in rough on the exam paper so you know what you are going to include and how you are going to structure the information. This should be done quickly; if you spend more than a minute on the notes it will probably be counter-productive.

The underlined words are important scientific terms and phrases and most, if not all, will be required. They will also need to be used correctly, often more than once.

Points to include:

- The principle that the <u>position of the equilibrium</u> will <u>shift</u> to <u>minimise the effect</u> of any imposed change in conditions.
- To maximise the percentage yield, the equilibrium needs to shift to the right.
- It is <u>an exothermic reaction</u>, so a low temperature shifts the equilibrium to the right. You know it is exothermic because, next to the equation $\Delta H_c^\ominus$ is negative.
- The imposed condition is a decrease in temperature. To minimise this imposed change the equilibrium moves right to release energy.
- There are nine moles of <u>gaseous reactants</u> and 10 moles of <u>gaseous products</u>.
- A low pressure shifts the equilibrium to the right because there are <u>fewer moles of products</u>.
- The equilibrium changes to minimise the effect of decreasing pressure by reducing the volume and making fewer gas molecules.

- An <u>excess</u> of oxygen/air will shift the equilibrium to the right so as to minimise the increase in concentration of oxygen.

A low level answer may have text that is not easy to read and punctuation and spellings that are not accurate. The information will not be organised very well and the style of writing will not show an examiner that you have understood a complex subject. The answer will not address all the points, or it will have some concepts wrong.

In a medium level answer the text is reasonably clear and some of the points are made but there is a lack of clarity. This will particularly apply to the concept of minimising the effect of an imposed change, which will probably be missing altogether.

A high level answer will be well organised and will include all the points above. Scientific vocabulary will be spelled correctly and used in the correct context and the answer will be grammatically sound.

A11 A structural model for benzene

1. 140 picometres could be written as 140×10^{-12} metres and then converted to standard form by moving the decimal point two more places to the left.

This gives 1.40×10^{-10} metres in standard form.

2. a) See Fig 86.

   b) There are three double bonds:
   $120 \text{ kJ mol}^{-1} \times 3 = -360 \text{ kJ mol}^{-1}$.

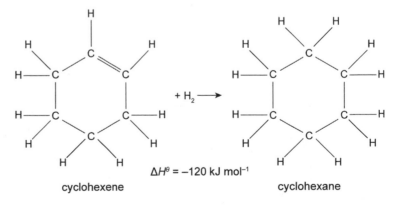

$\Delta H^\ominus = -120 \text{ kJ mol}^{-1}$

cyclohexene   cyclohexane

△ Fig 86 Answer to Q 2a).

3. a) See Fig 87.

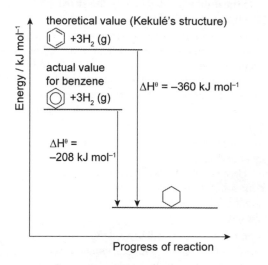

△ Fig 87 Answer to Q 3a).

b) Enthalpy change of hydrogenation of Kekulé's structure is −360 kJ mol⁻¹. This starts at a higher energy level in the diagram so it must be less stable than the actual benzene molecule, which is at a lower energy level.

4. **QWC** For a model to be accepted it must explain all known observations.

- Kekulé's model did not explain why the enthalpy change of hydrogenation was lower than expected.

- Different analytical tools were developed. X-ray crystallography and infrared spectroscopy results about bond lengths could not be explained by Kekulé's model.

- So anomalous results built up and a new model was needed.

- When Kekulé proposed his model no one knew about atomic orbitals but scientists could use this new understanding in developing a new model.

- The new model did explain the anomalous results.

A high level response will use the term *model* in the correct context and will contain the first bullet point and the last as well as at least two other points. The structure will show good organisation.

A medium level answer may not explain the significance of the first point and the last point will probably be missing. There will be reasonable organisation of the answer but some of the other points above will not be explained clearly.

A low level answer may well be characterised by a lack of clarity and organisation. The first and last points will be missing or poorly described.

A12 TNT – a formidable explosive

1. See Fig 88.

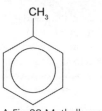

△ Fig 88 Methylbenzene

2. a) For this equation it is helpful to use the molecular formula for TNT so you can easily see how many atoms of the different elements you have to balance.

$$2C_7H_5N_3O_6(s) \rightarrow 3N_2(g) + 5H_2O(g) + 7CO(g) + 7\,C(s)$$

b) From the equation:
2 moles of TNT gives 15 moles of gases.

So 1 mole of TNT gives 7.5 moles of gases.

Volume of 1 mole of any gas at 298 K and 1 kPa = 24 dm³

Volume of 7.5 moles = 24 × 7.5 = 180 dm³

Even if you do not write a correct equation the examiner will calculate the volume that would be produced from your equation. This is called *error carried forward*. Even if you suspect you have the equation wrong, do not cross it out.

c) $M_r\,C_7H_5N_3O_6 = (12.0 \times 7) + (1.0 \times 5) + (14.0 \times 3) + (16.0 \times 6) = 227.0$

$$\text{moles of TNT} = \frac{\text{mass in grams}}{\text{mass of one mole (in grams)}}$$

$$= \frac{1\,g}{227\,g} = 0.004\,405 \text{ moles}$$

From equation: 1 mole of TNT gives 7.5 moles gases

So 4.405 × 10⁻³ moles gives:
4.405 × 10⁻³ x 7.5 = 0.033 04 moles

Volume of gas produced from 1 g = 24 000 × 0.033 04 = 793 cm³ to 3 sf

d) The black smoke indicates the formation of carbon.

3. Activation energy is the minimum energy with which two particles (atoms, molecules or ions) must collide for a reaction to occur.

4. a) Remember this reaction does not occur, but if it did the bromine would add across the double C=C bonds (see Fig 89).

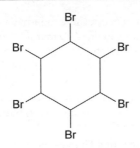

△ Fig 89 Answer to Q 4a).

As bromine water is used, another possibility is up to three OH groups could attach instead of Br atoms (see Fig 90).

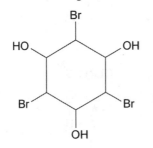

△ Fig 90 OH groups attaching instead of Br atoms.

b) See Fig 91.

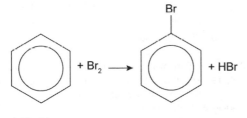

△ Fig 91

You could also write: $C_6H_6 + Br_2 \rightarrow C_6H_5Br + HBr$.

This reaction requires a catalyst such as $AlCl_3$ or $AlBr_3$ but this need not be shown in the balanced equation.

5. **QWC** The aspects that will require particular attention in this question are the accuracy of your reaction mechanism and the clarity of your explanation of the term *electrophile*.

* Reagents and conditions are concentrated HNO_3 and concentrated H_2SO_4 at 30 °C.

* The electrophile is NO_2^+.

* An electrophile is a particle that can accept a lone pair of electrons to form a covalent bond.

All three points are asterisked because each one will gain a mark.

High level responses must have all three.

Medium level responses may well have the first two but lack clarity in the third point. This is a definition you should learn carefully. Words like *electron-loving* lack the clarity of meaning required.

A low level response may have the temperature incorrect and will not cover the third point very well.

Now let's look at the mechanism. This does require a very clear, accurate answer and the minimum for a high level response is shown in Fig 92.

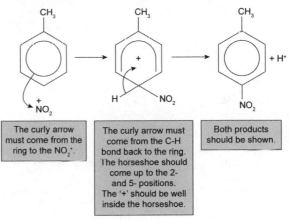

| The curly arrow must come from the ring to the NO_2^+. | The curly arrow must come from the C-H bond back to the ring. The horseshoe should come up to the 2- and 5- positions. The '+' should be well inside the horseshoe. | Both products should be shown. |

△ Fig 92 The reaction mechanism for the electrophilic substitution of methylbenzene.

The curly arrow represents the direction of movement of a lone pair of electrons so it needs to be precisely drawn each time. The information in the boxes tells you how you must draw the mechanism for a high level response.

A medium level response may lack some of the clarity required. For example, one of the curly arrows may not clearly start where it does on the diagrams above.

A low level response may well have the direction of the curly arrows incorrect or miss out the positive charge somewhere.

A13 Peppermint in medicine

1. **a)** A functional group is the atom, or group of atoms, that determines the chemical properties of a molecule.

   b) Menthol: alcohol group.
   Menthone: ketone group.

2. First work out the molecular formula for each molecule. This is best done by changing the skeletal formulae into a structure where you can easily count the number of atoms of each element (see Fig 93).

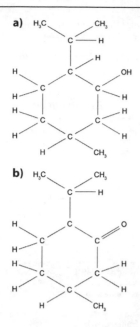

a) H₃C, CH₃

△ Fig 93 Structure of (a) menthol and (b) menthone.

a) Menthol

$C_{10}H_{20}O$ $M_r = (12.0 \times 10) + (1.0 \times 20) + 16.0$

$= 156$

b) Menthone

$C_{10}H_{18}O$ $M_r = (12.0 \times 10) + (1.0 \times 18) + 16.0$

$= 154$

3. **a)** 2,4-dinitrophenylhydrazine (DNPH) reacts with carbonyl groups (C=O) to give orange precipitates. Since menthone is a ketone it gives an orange precipitate with DNPH. Menthol will show no change with DNPH.

b) Tollens' reagent is used to test for aldehyde functional groups (–CHO). Since neither of these compounds is an aldehyde they will not give a positive result of a *silver mirror*.

4. **a)** This is an oxidation reaction, so acidified potassium dichromate solution will be the oxidising reagent and the reaction requires heating.

   • The alcohol group involved is a secondary alcohol (>CHOH) so a ketone is produced.

   b) This is a reduction, so NaBH₄ would be a good reagent to use.

5. • Gas chromatography is a very good technique to use for separating complex mixtures such as that of peppermint oil.

   • Once the compounds have been separated they can be individually analysed by a mass spectrometer and identified.

   • Separation is required because several molecules in the mixture may have very similar relative molecular masses and fragmentation patterns in the mass spectrometer, making accurate identification impossible.

6. **QWC** The explanation of optical isomerism needs to be clear and you should draw 3D mirror image diagrams. Remember that the question asks you to use to use one of these compounds as an example.

   • Optical isomerism arises when molecules have the same structural formula (when atoms in a molecule are bonded in the same order) but they are arranged differently in space.

   • Optical isomers are formed if there are four different groups bonded to a carbon atom.

   • These carbon atoms have a chiral centre, which means they have no plane of symmetry.

   • The optical isomers are non-superimposable mirror images of each other.

   • We have chosen one part of menthol to illustrate this, in Fig 94:

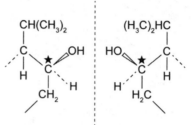

imaginary mirror

⟋ bond going out of the plane of the paper

⟋ bond going in to the plane of the paper

⟋ bond on the plane of the paper

△ Fig 94 Part of the menthol molecule showing a chiral carbon and the mirror image. You could use any of the asymmetric carbons. It is often helpful to include an imaginary mirror.

Although the chiral carbon is bonded to two carbon atoms, these atoms are part of two different groups.

Now you can identify the chiral carbons in each of the molecules. These are usually indicated by a star, as in Fig 95.

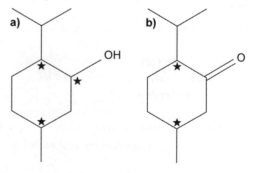

△ Fig 95 Chiral carbons in a) menthol and b) menthone.

High level responses will explain optical isomerism using drawings of one of the structures in 3D. All the bullet points will be included and all the chiral carbons will be clearly shown in both molecules.

Medium level responses will have some of the bullet points but may not use one of the molecules in the question as an example. If the example is used then it may not be clear and the molecule may not have the 3D convention for bonds around the chiral carbon. Probably all the chiral carbons will be identified.

A low level response will not explain optical isomerism clearly and is unlikely to draw a structure of menthol or menthone. It will probably look disorganised and be difficult to follow. The identification of some of the chiral centres may be incorrect, or some will be missed.

A14 Saturated and unsaturated fatty acids

1. **a)** The systematic name for glycerol is propane-1,2,3-triol.

   b) The functional groups are alkene and carboxylic acid. Remember you are asked to *name* them.

2. **a)** Systematic names are agreed by an organisation called IUPAC (International Union of Pure and Applied Chemistry). As it is international, chemists everywhere can communicate with clarity about the compounds they are using in experiments so that everyone understands exactly what these are.

   b) • The first number indicates the number of carbon atoms.

   • The second number tells you how many double bonds are present.

   • The number in brackets gives you the number of the carbon atom to which the double bond is attached, numbering the carbon of the carboxylic group as number 1.

   • Linoleic acid has 18 carbon atoms and 2 double bonds. One of these double bonds is attached to carbon atom 9 and the other to carbon atom 12. You will see this more clearly if you draw out the structure.

3. **a)** See Fig 96.

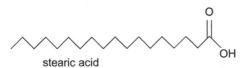

stearic acid

Δ Fig 96 Skeletal formula of stearic acid.

   b) Stearic acid, $CH_3(CH_2)_{16}COOH$:

   Molecular formula is $C_{18}H_{36}O_2$:

   $M_r = (12.0 \times 18) + (1.0 \times 36) + (16.0 \times 2) = 284.0$

   Oleic acid, $CH_3(CH_2)_7CH = CH(CH_2)_7COOH$:

Molecular formula is $C_{18}H_{34}O_2$:

$M_r = (12.0 \times 18) + (1.0 \times 34) + (16.0 \times 2) = 282.0$

   c) There is less <u>surface contact</u> between the oleic acid molecules than there is between the stearic acid molecules because the oleic acid molecules are bent. This means there are <u>fewer van der Waals forces</u> acting between oleic acid molecules. Hence oleic acid has a lower melting point. (The underlined terms are essential.)

4. **a)** See Fig 97.

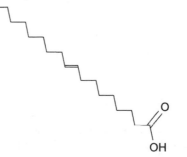

Δ Fig 97 *trans*-oleic acid (*E*-oleic acid).

   b) Enzyme active sites will not be able to accept the differently shaped *trans* fat, whereas *cis*-fatty acids will fit and can be metabolised.

5. **a)** A nickel catalyst is used to hydrogenate vegetable oils.

   b) • The oil molecules and the hydrogen molecules are <u>adsorbed</u> onto the surface of the catalyst.

   • This causes the carbon-carbon double bonds and the H-H bonds to weaken, which lowers the activation energy.

   • The hydrogenated fat molecule then <u>desorbs</u> away from the catalyst surface.

   The underlined terms are essential.

6. There must be an excess of hydrogen so that all the double bonds react.

   1 mole of H_2 reacts with each double bond in a molecule, so you could analyse a hydrogenated sample with a mass spectrometer and for every 2.0 the molecule has increased in M_r there would be a double bond (see Fig 98).

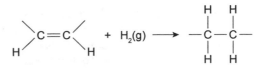

Δ Fig 98 One mole of hydrogen adds across one C-C double bond.

7. Drawing out the molecule will help you work out which carboxylic acids may be formed when the C-C double bonds are split.

   $CH_3(CH_2)_4CH=CHCH_2CH=CHCH_2(CH_2)_6COOH$

Two possible carboxylic acid molecules are shown in Fig 99:

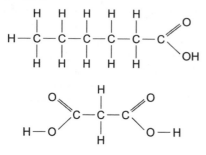

△ Fig 99 Two carboxylic acids that may form from the oxidative splitting of the C-C double bonds.

Any carboxylic acid derived from the structure of linoleic acid would be acceptable.

8. **QWC** In this question, your audience is the general public. It should be a balanced article that will need to explain clearly the key scientific terms. In the list below, the most important points have more asterisks.

**Briefly explain what is meant by unsaturated fatty acid, *trans* fatty acid, fats and oils.

**Explain why unsaturated fats are hydrogenated, including a meaning of this term.

**At first saturated fats were thought to be linked to heart disease so sales of low fat spreads containing unsaturated fats increased.

* Scientific studies began to suggest a correlation between *trans* fatty acids and increased risk of heart disease.

**Some scientists suggested a causal link. This is when one factor is found to directly affect another.

***Scientists also suggested a plausible mechanism to explain the link, which meant that it was much more readily accepted by the scientific community.

***The mechanism is that *trans* fatty acids are not metabolised by enzymes in the body because they do not fit into the enzymes' active sites. This could be the reason why *bad* cholesterol increases in the blood.

* *trans* fats were also linked to an increase in obesity.

* Information about *trans* fatty acids was explained to the public and government.

* Governments acted by passing laws insisting that the proportion of *trans* fats was clearly labelled on processed foods.

* The public acted by not buying these products.

* Food manufacturers found alternatives to reduce or eliminate *trans* fats.

**The causal link between *trans* fats and health risks is still disputed.

***Your conclusion about whether society should have made the decisions it did.

There are a large number of points here.

High level answers

- Explain at least three scientific terms clearly to the newspaper audience.
- They will use at least three of the underlined terms and explain these.
- Six points including the last one will be present and these will be ones with more asterisks.

Medium level responses

- Explain fewer scientific terms and fewer underlined words.
- The organisation of the answer will not be as good.
- Fewer points, with fewer asterisks, may be used or they will not be explained with clarity.
- The last point may not be mentioned.

Low level responses

- May be disorganised.
- Fail to explain clearly the scientific terms.
- Use points with fewer asterisks.
- The answer will not be written for its intended audience.

A15 Aramids: fire-resistant and bulletproof

1. **a)** See Fig 100.

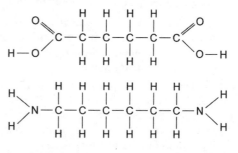

△ Fig 100 The two monomer molecules that make up the repeat unit of nylon-6,6.

b) The molecular formula of the repeat unit is $C_{12}H_{22}N_2O_2$.

$M_r = (12.0 \times 12) + (1.0 \times 22) + (14.0 \times 2) + (16.0 \times 2) = 226$

Average M_r of polymer $= 1.13 \times 10^5$, so number of repeat units $= \dfrac{1.13 \times 10^5}{226} = 500$

c) The polymer chains will not all be the same length so the number of repeat units will vary.

2. a) i) acyl chloride; **ii)** amine.

b) See Fig 101.

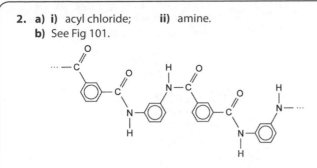

∧ Fig 101 Nomex®, showing two repeat units.

c) Secondary amide functional group, or just amide functional group.

The functional group here is not a peptide because it does not join two amino acids. This is a common mistake.

d) Condensation polymer.

3. See Fig 102.

(a) 1,4-diaminobenzene

(b) 1,4-benzenedicarboxylic acid

∆ Fig 102 (a) 1,4-diaminobenzene and (b) 1,4-benzenedicarboxylic acid.

b) See Fig 103.

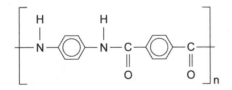

∆ Fig 103 Repeat unit of Kevlar®.

- When drawing the repeat unit, the end bonds must cut the square brackets.
- You should write n next to the last bracket because the number of repeat units in a chain will vary.

4. **QWC** Points for your answer include:

- The intermolecular force is the hydrogen bond.
- Hydrogen bonds form when hydrogen is covalently bonded to the very electronegative elements oxygen, nitrogen and fluorine.

- The partial positive charge on the hydrogen is attracted to the lone pair on the oxygen.
- Drawing out a small section of polymer is a good way to show how they form (see Fig 104).

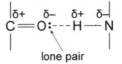

∆ Fig 104 Formation of a hydrogen bond.

Notice that we have included the lone pair of electrons, which is really an essential part of the hydrogen bond. You will sometimes see mark schemes that do not require this but you cannot lose a mark if you include it.

- The hydrogen bonding is stronger in Kevlar® because the polymer chains are able to get closer together as they are linear.
- Nomex® has a lower melting point because the chains are not linear, so parts of the molecule are not close enough for as many hydrogen bonds to form.

A high level response will include all six points, clearly shown, in a logical order.

A medium level response may miss drawing the diagram and will not clearly make the last two points.

A low level response will be difficult for an examiner to follow. It may mention hydrogen bonding, but there is unlikely to be a diagram.

The explanation about strength of bonding will probably not show the significance of the linear chains in Kevlar®.

A16 Reaction kinetics and vehicle exhausts

1. a) First order because the half-life is constant.

b) If you put 4.24×10^{-4} into your calculator and divide it by 2, and repeat this two more times, you get three half-lives which reach 5.30×10^{-5} mol dm^{-3}.

$3.8 \times 3 = 11.4$

11 hours to 2 sf.

2. a) Third order (add the powers of the concentrations together).

b) Rate = $k[NO(g)]^2[O_2(g)]$

You need to change the subject of the equation.

$$k = \frac{rate}{[NO(g)]^2[O_2(g)]}$$

$$k = \frac{1.60 \times 10^{-3}}{(2.00 \times 10^{-3})^2 \times 3.00 \times 10^{-3}}$$

Units of $k = \dfrac{\text{mol dm}^{-3}\,\text{s}^{-1}}{(\text{mol dm}^{-3})^2 \times (\text{mol dm}^{-3})}$

$k = 1.33 \times 10^5\,\text{dm}^6\,\text{mol}^{-2}\,\text{s}^{-1}$ to 3 sf

Notice that the units start with dm^6 because it has a positive power.

c) i)   The rate increase by four times.

   ii) The rate decreases by a half.

   iii) The rate increases by $(4)^2 \times 4 = 64$ times

d) **First step** $2NO(g) + O_2(g) \rightarrow N_2O(g) + O_3(g)$ *slow*

   Second step $N_2O(g) + O_3(g) \rightarrow 2NO_2(g)$ *fast*

- You would get one mark for understanding that the slower step must involve the collision of one possibility of NO and one molecule of O_2 because this is in the experimentally determined rate equation.

- There is no clue in the question about the products of the slow step, so you can suggest any that seem sensible to you. We have given one possibility but you may think of others.

- The second step, when added to the first step, must produce the overall equation and this would give you your second mark.

3. **a)**

- Comparing experiments 1 and 2, the concentration of $NO_2(g)$ doubles, which doubles the rate of reaction.

  The concentration of $CO(g)$ is constant.

  Therefore: rate $\propto [NO_2(g)]$.

  So, the reaction is <u>first order</u> with respect to $NO_2(g)$.

- Comparing experiments 1 and 3 the concentration of $CO(g)$ doubles, which doubles the rate of reaction.

  The concentration of $NO_2(g)$ is constant.

  Therefore: rate $\propto [CO(g)]$.

  So, the reaction is <u>first order</u> with respect to $CO(g)$.

b) Since rate $\propto [NO_2(g)][CO(g)]$ the rate equation is:

   rate $= k[NO_2(g)][CO(g)]$

c) The overall order of reaction is the sum of the powers in the experimentally determined rate equation.

   This is $1+1 = 2$

   ($[NO_2(g)]$ and $[CO(g)]$ are both to the power 1.)

   So the reaction is overall second order.

d) $k = \dfrac{\text{rate}}{[NO_2(g)][CO(g)]}$

   Units of $k = \dfrac{\text{mol dm}^{-3}\,\text{s}^{-1}}{(\text{mol dm}^{-3}) \times (\text{mol dm}^{-3})}$

Now substitute into the equation one of the sets of values. For example, take experiment 1.

$k = \dfrac{6.40 \times 10^{-6}}{4.80 \times 10^{-1} \times 2.52 \times 10^{-2}}$

Units of $k = \text{dm}^3\,\text{mol}^{-1}\,\text{s}^{-1}$

Therefore:

$k = 5.29 \times 10^{-4}\,\text{dm}^3\,\text{mol}^{-1}\,\text{s}^{-1}$ to 3 sf

4. **QWC** The way you organise your answer to explain clearly which mechanism best agrees with the experimental evidence will make the difference between high and medium level answers to this question.

- Both mechanisms have steps that when added together give the overall equation:

  $NO_2(g) + CO(g) \rightarrow NO(g) + CO_2(g)$

- The reaction is zero order with respect to $CO(g)$, so it will not be present in the rate determining step.

- The rate determining step is the slow step.

- Neither of the slow steps has any CO in it. This is consistent with both mechanisms.

- The reaction is second order with respect to $NO_2(g)$, so the slow step must involve a collision between two molecules of $NO_2(g)$.

- Mechanism 1 best agrees with the experimental evidence.

High level answers will make all the points above and will link the slow step to the collision of the two molecules of $NO_2(g)$.

Medium level answers will make most of the points but will not use all the evidence in the question, such as the relevance of zero order with respect to $CO(g)$.

Low level responses will lack organisation and may not mention orders of reaction, or rate determining steps. Very few of the points above will be clearly made.

AIS1

1. **a)** See Table 19.

Titration number	1	2	3	4
Final burette reading/cm³	21.50	42.20	21.35	42.15
Initial burette reading/cm³	0.50	21.40	0.80	21.50
Titre *or* Volume added/cm³	21.00	20.80	20.55	20.65

△ Table 19

b) The mean from the two closest titres (20.55 cm³ and 20.65 cm³) is 20.60 cm³.

c) 0.002 06 mol NaOH

d) 0.002 06 mol sulfamic acid

e) 0.0206 mol sulfamic acid

f) 0.824 mol dm^{-3} sulfamic acid

g) i) ± 0.05 cm^3

   ii) $\dfrac{2 \times .05}{v} \times 100$

   (where v is one of the titres in Table 19)

h) i) To ensure that the concentration of any sulfamic acid solution in the pipette is the same as the solution to be pipetted.

   ii) Any sulfamic acid solution remaining in the pipette before it is used to deliver 25 cm^3 of the diluted solution will be more concentrated. This will mean there are more moles than expected so the titre of sodium hydroxide solution will be greater.

i) No effect because the number of moles of sulfamic acid delivered by the pipette are the same.

j) The air bubble disappeared during the experiment so it will appear that the volume, and therefore the amount, of NaOH(aq) is more than expected to neutralise the sulfamic acid.

2. a) 1.70 g (*not* 1.7 g)

   b) $Na_2CO_3(aq) + 2HA(s) \rightarrow 2NaA(aq) + CO_2(g) + H_2O(l)$

   c) 0.008 625 mol carbon dioxide

   d) 0.017 25 mol sulfamic acid

   e) Molar mass of sulfamic acid = 98.6 g

   f) 0.483%

   g) Either the bung was not replaced quickly enough at the start of the reaction so some CO$_2$ escaped *or* some CO$_2$ dissolved in the water.

   h) Repeat the experiment and take the most precise readings.

3. a) Sodium hydroxide is corrosive and so safety goggles should be worn. Other possible answers relating to the corrosive nature of NaOH could be to wear plastic gloves or to mop up any spillages with plenty of water.

   b) Neutralise sulfamic acid with a base before tipping down the sink.

4. 49.43% is oxygen and the molecular formula of sulfamic acid is H_3NSO_3.

AIS2

1. a) See Fig 105.

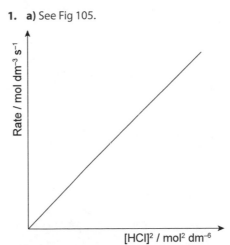

△ Fig 105 Graph of student's prediction.

- The diagonal line should pass through the origin.
- Remember you should have included the units.

b) i) Moles H$_2$ = $\dfrac{240}{24\ 000}$ = 0.010 mol. From equation moles of HCl will be 0.020 mol. 40 cm^3 of 0.50 mol dm^{-3} acid is required.

   ii) So that this variable is controlled and almost constant. Any form of words that shows your understanding that if magnesium is all used up before the hydrochloric acid then the experiment will not be valid.

c) There are many variations and Table 20 is one example. The units are included in the column headings.

Expt. no.	Volume of 0.50 mol dm^{-3} HCl(aq) /cm³	Volume of H$_2$O/ cm³	Concentration of HCl(aq)/ mol dm^{-3}
1	40	0	0.50
2	36	4	0.45
3	32	8	0.40
4	28	12	0.35
5	24	16	0.30

△ Table 20

ii) The independent variable is the <u>concentration</u> of HCl. If you put moles, volume or amount this would not receive any credit.

The dependent variable is time. This is very unusual as time is not normally the dependent variable. If you wrote rate this is also incorrect because that is a *derived* variable.

iii) These steps should be included:

- Start the stop clock at the same time as the reactants are mixed.
- Record volume of hydrogen produced at regular time intervals.
- Repeat the experiment using a different concentration of HCl.

d) i) See Fig 106.

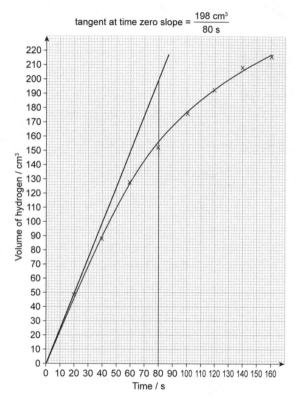

tangent at time zero slope $= \dfrac{198 \text{ cm}^3}{80 \text{ s}}$

△ Fig 106

ii) Initial rate of reaction $= \dfrac{198 \text{ cm}^3}{80 \text{ s}}$

$$= 2.5 \text{ cm}^3\text{ s}^{-1} \text{ to 2 sf.}$$

You would be allowed $\pm 0.5 \text{ cm}^3\text{ s}^{-1}$ and you would be expected to include the units.

2. a) See Table 21.

Time/min	[sucrose(aq)] /mol dm⁻³	log₁₀[sucrose(aq)]
0	0.500	−0.301
40	0.427	−0.370
80	0.363	−0.440
120	0.309	−0.510
160	0.288	−0.540
200	0.224	−0.650
240	0.191	−0.719
280	0.162	−0.790

△ Table 21

Note: In logarithm values it is the numbers to the right of the decimal point that are significant. See S32 and S35.

b) See Fig 107. This graph shows the answers to questions 2b), 2c) and 2e).

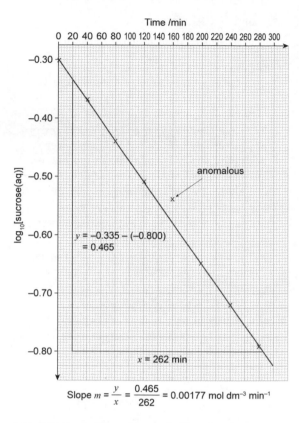

$y = -0.335 - (-0.800)$
$= 0.465$

$x = 262$ min

Slope $m = \dfrac{y}{x} = \dfrac{0.465}{262} = 0.00177 \text{ mol dm}^{-3} \text{ min}^{-1}$

△ Fig 107 Graph of log₁₀ [sucrose(aq)] against time, showing the anomalous point and calculation of the slope.

Points an examiner will check:

- Are both axes labelled and does time have a unit? Usually the log of a value is not considered to have a unit.
- The y-axis should have descending values for log₁₀[sucrose(aq)].
- Does the graph occupy at least half the available space on each axis?
- Are all eight points plotted correctly? In an examination this should normally be within half a small square.
- Does the line of best fit go through most of the non-anomalous points and, if not, are they evenly spaced above and below the line?
- Is the y-axis intercept plotted at (0, 0.005)?

c) i) One point is anomalous, shown in Fig 107.

You should explain that the concentration is higher than expected so:

- *either* the sample was removed too early
- *or* the recorded time is later than when the sample was actually taken.

ii) Very reliable because, with the exception of the anomalous point, they all lie close to, or on, the straight line.

d) At 100 min $\log_{10}$ [sucrose(aq)] = −0.475.
Therefore [sucrose(aq)] = 0.335 mol dm^{-3} to 3 sf.

e) See Fig 107.

   i) $\log_{10}$ [sucrose(aq)]
      = $-kt + \log_{10}$ [initial sucrose(aq)]

      y is $\log_{10}$ [sucrose(aq)]; m is $-k$; x is time;
      c is $\log_{10}$ [initial sucrose(aq)]

   ii) m is the slope, which is $-k$.
      $k = 0.001\ 77$ mol dm^{-3} min^{-1}

f) Yes, because a straight line graph has been produced.

g) i) Temperature is a variable that must be controlled for the results to be **valid.** Rate is affected by temperature so if the temperature changes the concentration of sucrose sampled at a particular time will change.

   ii) You should have drawn a graph with a shallower gradient, which should intercept the y-axis at 0.301 ± half a small square.

      This is because a lower temperature produces a slower rate of reaction so the concentrations of the samples taken at the times specified will be less.

QWC Worked Examples

At some point during your course, you will be assessed on the quality of your written communication. These annotated worked examples show how low, medium and high scoring responses to Question 3 in Activity 3 gain the marks they do.

3. The first two ionisation energies of caesium and barium are shown in Table 7. Explain this data in terms of the electron configuration of these atoms. **QWC**

Element	First ionisation energy/kJ mol−1	Second ionisation energy/kJ mol−1
caesium	376	2234
barium	503	965

△ Table 7

★ Low level answer

This is a weak answer. It has poor organisation and little structure. A few moments spent thinking about what the question was asking would have really helped here. Before you start an answer it is a good idea to jot down the points that occur to you. You can put a line through this later to tell the examiner that you do not want it marked.

This does not tell you which element is in which group.

Misspelling words that are in the question is poor – "caesium" and "energy" are spelled incorrectly. Energy is an important scientific term.

There should be a full stop here followed by a capital letter for a new sentence.

This does not mention how large the difference in ionisation energies is.

The student's answer has no slang words and the writing is quite formal, which are points in its favour – showing it is written for its intended audience. In this case the only audience you are asked to write for is the examiner marking the question.

> *These elements are in group 1 and 2. Cesium has a difference in enegy so there is one electron in the outer shell there is also a difference in enegy for barium so these electrons are in the same shell and there are two.*

The answer could be rewritten in the following way to make the meaning clearer.

Caesium is in Group 1 and barium is in Group 2. There is a large difference in ionisation energy when the second electron is removed from caesium. There is a smaller difference in ionisation energies in barium so these electrons are in the same shell.

Three *clear* points are now made:

- The positions of caesium and barium in the periodic table.
- The large jump in energy when the second electron is removed from caesium and a smaller difference when this electron is removed from barium.
- The two electrons in barium are in the same shell.

★★ Medium level answer

This answer is more organised because it shows a logical progression in the points that are made. The significance of the positions of the two elements in the periodic table is not mentioned. Under the pressure of the examination this does happen so read the question carefully to make sure you have answered all the points. There are no spelling, grammar or punctuation errors and the meaning of what is written is clearer than the low level answer.

First and second ionisation energies are chemical terms that are used appropriately.

There is a big jump in the first and second ionisation energies of caesium. This shows that the first electron is in the outer shell. The second electron comes from the next full shell. The ionisation energies of barium are close together so these electrons are in the first shell.

Caesium is in Group 1 and barium is in Group 2.

This is incorrect because the first shell is the one closest to the nucleus.

There is no mention of the significance of Groups 1 and 2 to the electron structure of these elements. The answer should say that because caesium is in Group 1 there will be one electron in the outer shell.

This high level answer is well written, with accurate punctuation and grammar, therefore the meaning is very clear. It has been well planned because all the important points are covered and is also well organised because the points are made in a logical progression. Chemical terms (specialist vocabulary) are used accurately so first and second ionisation energies are explained. The student has thought about the intended audience, which in this case is the examiner. The writing is formal and appropriate. This answer would gain you full marks.

From the periodic table caesium is in Group 1 and barium is in Group 2. This means that caesium has one electron in its outer shell and barium has two.

This explains the significance of the positions of the elements in Groups 1 and 2.

The first ionisation energy is the energy required to remove 1 mole of electrons from 1 mole of gaseous atoms.

$E(g) \rightarrow E^+(g) + e^-$ *where E is any element.*

The second ionisation energy is the energy to remove a second mole of electrons:

$E^+(g) \rightarrow E^{2+}(g) + e^-$

The question is part of an activity that defines first and second ionisation energies, so there is no need to re-define them here. However, when answering a question where no definitions are present, you would be expected to define what these terms mean, so we have included them here.

For each element, ionisation energy increases as each successive mole of electrons is removed. This is because the positive nuclear charge remains the same but the number of electrons is decreasing.

This point is often missed but it is still important to explain why successive ionisation energies increase.

In caesium there is a large jump in energy when the second mole of electrons is removed. These electrons are coming from a full shell of electrons which is closer to the nucleus. This is evidence for only one electron in the outermost shell. This electron is also less shielded from the positive nuclear charge because there are fewer inner electron shells.

This is a very clear explanation of why there is a large jump in energy.

This is a very well-organised answer because it shows why this is evidence for the electron structure of these atoms.

Glossary

accuracy How close a measurement is to its true or accepted value.

amount The number of particles in a substance. It has the unit moles (mol).

anomalous result An unexpected result that does not fit with the currently accepted theory.

aqueous solution A solution where the solvent is water. In equations the state symbol of a dissolved substance is (aq).

bias When results are shifted in a particular direction to support a hypothesis.

causal link When one factor (variable) directly affects another.

concentration The mass or amount of substance in a certain volume.

conclusions These are the judgments made about what the data means.

continuous data This is data (results) where each value can be any number between two limits.

correlation When one variable appears to have a relationship with another, so that when one changes so does the other.

data Observations, such as measurements, that are collected, often in an experiment.

decimal places The number the digits, including zeros, to the right of the decimal point.

dependent variable This is when the value of a measurement changes and this change is caused by something else changing (the **independent variable**).

directly proportional ($\propto$) This means that as one variable increases, the other increases by the same percentage.

discrete data The values in the data are separate and can only be particular numbers.

displayed formula A formula that shows all the atoms and bonds in a molecule.

electron pair repulsion model This is a model for determining the shapes of molecules using the repulsion of electron pairs around a central atom.

end point In volumetric analysis this is when one reactant exactly reacts with another.

estimating Doing very rough calculations, often using rounded-up numbers to one significant figure.

ethics A framework to guide how we make decisions by considering rightness and wrongness of actions.

formula (formula unit) If a compound is made up of ions rather than molecules, the formula tells you the ratio of the number of each type of ion present in the structure. For example, in NaCl there are equal numbers of sodium and chloride ions. (See also **molecular formula**.)

fraction A fraction tells you how much of something there is, shown as a line between two numbers. The number on top (or alongside) the line is the actual number of parts and the number below (or after) the line is the total number of possible parts.

hazard Anything that can cause harm.

human errors In experiments these are errors made by people taking the measurements.

hypothesis An idea that seeks to explain what is causing something to occur as it does. Predictions are made from a hypothesis and these can be tested experimentally.

independent variable Changing measurements that are not linked to something else changing.

isotopes Atoms of the same element, so they have the same atomic number (proton number) but different mass numbers.

line of best fit This is a line drawn to go through plotted points, so that most lie on the line or are roughly evenly spaced either side of the line.

logarithm (log) This is the exponent (power) to which a number must be raised. In your chemistry course the number is likely to be 10 and we call this number *base 10*. Logs reduce a very wide range of values to a more manageable scale.

mass number The total number of protons and neutrons in an atom.

mass spectrometers Instruments used to calculate relative atomic and molecular masses very accurately.

mean This is the average of a set of values. The mean of a set of repeated results will often produce a more accurate value.

methodology The procedures, equipment and techniques that have been used to produce the data.

model This is another way of explaining observations. It often simplifies quite complex ideas in a way that makes them easier to understand.

molar gas volume This is the volume of one mole of any gas. At 298 K and 101 kPa (sometimes called room temperature and pressure) the molar gas volume is 24 dm³.

molecular formula Tells you exactly how many atoms of each element there are in a molecule.

moles One mole is the amount of substance that contains as many particles as there are atoms in exactly 12 g of carbon-12.

One mole of substance is 6.02×10^{23} particles.

neutralisation point In acid-base titrations this is where the acid exactly neutralises the base.

peer review The checking of scientific papers or conference proposals by other scientists (the peers), who are experts in a similar field of research.

percentage (%) This means 'per hundred', so is a number expressed as a fraction of 100.

precise How closely measurements are clustered. If measurements are close together then they are said to be precise. The closer the grouping of results the more precise they are.

predictions These are statements of what you think will happen. They can be made from a hypothesis and then experiments can be carried out to see if these predictions are correct.

qualitative data Observations you can describe but not measure.

quantitative data These are observations you can measure.

random errors Errors in measurements that have no pattern.

rate equation This states the relationship between the rate of reaction and the concentration of each reactant. It can only be derived by experiment.

ratio A way of comparing two quantities and is usually written as x:y

relative atomic mass, A_r The average mass of an atom (taking into account all its isotopes and their abundance) compared with 1/12 the mass of one atom of carbon-12.

relative formula mass, M_r When a compound is made of ions it is more correct to use this term rather than relative molecular mass. (See **relative molecular mass**.)

relative molecular mass, M_r Calculated using the relative atomic mass scale, with carbon-12 as the standard against which the masses of molecules or formula units are compared.

reliable Results that can be repeated or reproduced.

repeated data When the same experimenter, using the same equipment, gets similar results.

resolution How well the apparatus detects small differences in measurement.

reproducible data Means that other scientists can produce similar results using the same methodology.

risk This is the chance of something harmful happening as a result of a hazard.

rounding The process by which you reduce the number of significant figures in your answer.

sensitivity Means the same as resolution. (See **resolution**.)

significant figures (sf) The number of digits from left to right in a particular value starting with the first digit that is not zero. It is a scientist's way of telling you how precise measurements are.

solute The substance that is dissolved in a solution.

solution What is produced when the solute dissolves in the solvent.

solvent The liquid that does the dissolving.

standard form Allows you to write very large or very small numbers using powers of 10.

state symbols These tell you what the physical states of the reactants and products are, usually at room temperature.

subject of an equation The variable or value in a mathematical equation that stands on its own next to the 'equals' sign

systematic errors These are errors that produce inaccurate values that are consistently higher or lower than expected.

tangent A line that just touches a circle at a point. It is used in reaction kinetics to find the slope of a curve at any point.

theoretical yield The maximum amount of products that are predicted to be obtained from a chemical equation.

theory When a hypothesis makes correct predictions it is called a theory. It often includes more than one hypothesis and explains a number of observations.

titration An experimental technique used by chemists to find out the amounts of substances that are reacting in known volumes of solution very accurately. This is also known as **volumetric analysis**.

valid When something does what it is meant to do. Results are valid when they measure what they are supposed to be measuring.

values These are numbers with meanings. The meanings are usually units of measurements, such as 298 K or 1 m.

variable Something that changes and has different values. (See **values**.)

volumetric analysis See **titration**.

Periodic table

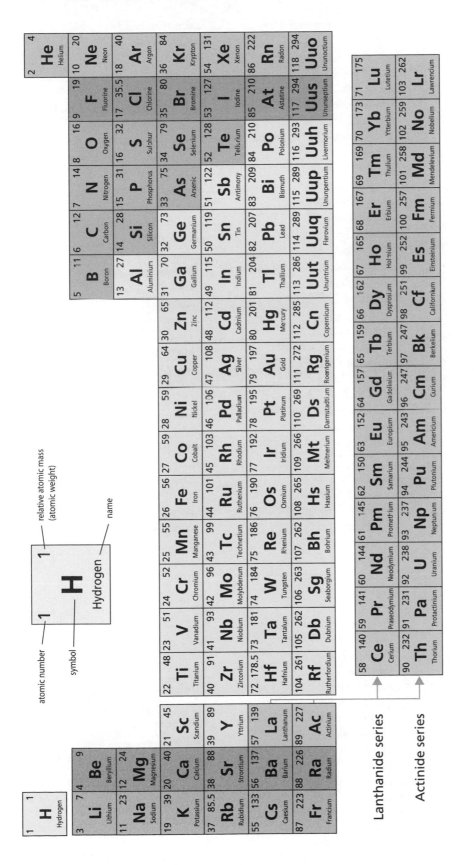

1	1
	H
	Hydrogen

atomic number — 1 — 1 — relative atomic mass (atomic weight)

symbol — **H** — Hydrogen — name

1	1
	H
	Hydrogen

3	7		4	9
	Li			**Be**
	Lithium			Beryllium

11	23		12	24
	Na			**Mg**
	Sodium			Magnesium

19	39	20	40	21	45
	K		**Ca**		**Sc**
	Potassium		Calcium		Scandium

37	85.5	38	88	39	89
	Rb		**Sr**		**Y**
	Rubidium		Strontium		Yttrium

55	133	56	137	57	139
	Cs		**Ba**		**La**
	Caesium		Barium		Lanthanum

87	223	88	226	89	227
	Fr		**Ra**		**Ac**
	Francium		Radium		Actinium

Lanthanide series

Actinide series

22	48	23	51	24	52	25	55	26	56	27	59	28	59	29	64	30	65
	Ti		**V**		**Cr**		**Mn**		**Fe**		**Co**		**Ni**		**Cu**		**Zn**
	Titanium		Vanadium		Chromium		Manganese		Iron		Cobalt		Nickel		Copper		Zinc

40	91	41	93	42	96	43	99	44	101	45	103	46	106	47	108	48	112
	Zr		**Nb**		**Mo**		**Tc**		**Ru**		**Rh**		**Pd**		**Ag**		**Cd**
	Zirconium		Niobium		Molybdenum		Technetium		Ruthenium		Rhodium		Palladium		Silver		Cadmium

72	178.5	73	181	74	184	75	186	76	190	77	192	78	195	79	197	80	201
	Hf		**Ta**		**W**		**Re**		**Os**		**Ir**		**Pt**		**Au**		**Hg**
	Hafnium		Tantalum		Tungsten		Rhenium		Osmium		Iridium		Platinum		Gold		Mercury

104	261	105	262	106	263	107	262	108	265	109	266	110	269	111	272	112	285
	Rf		**Db**		**Sg**		**Bh**		**Hs**		**Mt**		**Ds**		**Rg**		**Cn**
	Rutherfordium		Dubnium		Seaborgium		Bohrium		Hassium		Meitnerium		Darmstadtium		Roentgenium		Copernicum

5	11	6	12	7	14	8	16	9	19	10	20
	B		**C**		**N**		**O**		**F**		**Ne**
	Boron		Carbon		Nitrogen		Oxygen		Fluorine		Neon

13	27	14	28	15	31	16	32	17	35.5	18	40
	Al		**Si**		**P**		**S**		**Cl**		**Ar**
	Aluminium		Silicon		Phosphorus		Sulphur		Chlorine		Argon

31	70	32	73	33	75	34	79	35	80	36	84
	Ga		**Ge**		**As**		**Se**		**Br**		**Kr**
	Gallium		Germanium		Arsenic		Selenium		Bromine		Krypton

49	115	50	119	51	122	52	128	53	127	54	131
	In		**Sn**		**Sb**		**Te**		**I**		**Xe**
	Indium		Tin		Antimony		Tellurium		Iodine		Xenon

81	204	82	207	83	209	84	210	85	210	86	222
	Tl		**Pb**		**Bi**		**Po**		**At**		**Rn**
	Thallium		Lead		Bismuth		Polonium		Astatine		Radon

113	286	114	289	115	289	116	293	117	294	118	294
	Uut		**Uuq**		**Uup**		**Uuh**		**Uus**		**Uuo**
	Ununtrium		Flerovium		Ununpentium		Livermorium		Ununseptium		Ununoctium

2	4
	He
	Helium

58	140	59	141	60	144	61	145	62	150	63	152	64	157	65	159	66	162	67	165	68	167	69	169	70	173	71	175
	Ce		**Pr**		**Nd**		**Pm**		**Sm**		**Eu**		**Gd**		**Tb**		**Dy**		**Ho**		**Er**		**Tm**		**Yb**		**Lu**
	Cerium		Praseodymium		Neodymium		Promethium		Samarium		Europium		Gadolinium		Terbium		Dysprosium		Holmium		Erbium		Thulium		Ytterbium		Lutetium

90	232	91	231	92	238	93	237	94	237	95	243	96	247	97	247	98	251	99	252	100	257	101	258	102	259	103	262
	Th		**Pa**		**U**		**Np**		**Pu**		**Am**		**Cm**		**Bk**		**Cf**		**Es**		**Fm**		**Md**		**No**		**Lr**
	Thorium		Protactinium		Uranium		Neptunium		Plutonium		Americium		Curium		Berkelium		Californium		Einsteinium		Fermium		Mendelevium		Nobelium		Lawrencium

Index

A

accuracy 7, 11
alcohol 45, 53
alkenes 52
alpha particles 10, 41
ammonia 48
ammonium nitrate 48
amounts 15, 17
animal fats 53
anomalous results 4–5, 8, 10
antilogarithm 27–8
aqueous solution 14, 20–1
aramids 55
arithmetic means 30
atomic mass 39, 40
atoms 4–5, 10, 14, 39–41, 42, 43, 47, 50
   electrons 42, 43, 44, 50
   emission spectra 42
   ionisation energies (IE) 41, 42
   isotopes 15
   nucleus 40, 41
   relative atomic mass 15, 25
   shells 41, 42, 43
   subatomic particles 40
   subshells 42
Avogadro, Amedeo 16, 18

B

bacteria 39
Bartlett, Neil 6, 43
benzene 4, 50–1, 55
beta particles 40
bias 12, 13
bioethanol 45
Brazil 45
breast cancer 5
bromine water 51
burettes 8, 9, 11
butane 19

C

calcium hydroxide 17
calculations
   antilogarithms 27–8
   calculators 27–8
   decimal places 30
   estimation 28–9
   fractions 26
   logarithms 27, 29
   order of magnitude 28
   percentages 26

   powers of numbers 28
   ratios 26
   significant figures 29–30
calculators 27–8
cancer 5, 52
carbon 46, 47, 50, 51
carbon-12 15, 25
carbon dioxide 8, 14, 16, 19, 45, 48
carbon monoxide 14, 51
carbon-neutral 45
carbon nanotubes 46
carboxylic acid 53
catalytic converters 14
causal links 9–10
CFCs (chlorofluorocarbons) 9–10
chlorine 10, 26
cholesterol 9, 53, 54
chromatography 6
cis-isomers (Z-fatty acids) 54
communicating scientific
   information 12–13
   bias 12, 13
   peer reviews 12
concentration 20, 34, 35, 52
continuous data 33
correlation 9

D

data 4, 5
   accuracy 7, 8, 10
   anomalous results 4–5, 8, 10, 11
   assessing error 11
   conclusions 11
   dependent variable 8–9
   evaluation 10
   graphs 32–3
   independent variable 8–9
   interpretation 9–10
   methodology 7, 8, 10
   precision 7, 10, 29
   qualitative 8
   quantitative 8, 32
   reliability of 4, 5–6, 10
   reproducibility 10, 11
   significant figures 29
   validity 5, 6, 11, 13
decimal places 24–5, 30
decimetres cubed (dm³) 18
decision-making 13–14
dependent variable 8–9

discrete data 33
double bonds 51, 54
drugs 5, 45
DuPont 55

E

electron pair repulsion model 36
electrons 4–5, 40, 41, 42
   bonding pairs 44
   lone pairs (non-bonding pairs) 44
electrophilic reactions 51
emission spectra 42
enthalpy changes 31
equations 14–22
   amounts 17–18, 31
   exponentials 24
   formulae 14, 16
   masses 17–18
   rate equations 35
   ratio 14
   standard form 24–6
   state symbols 14
   subjects of 31
   theoretical yield 17
   volumes 19–20
esters 33–5, 53
estimation 28–9
ethane 37
ethanol 45
ethics 13–14
experiments
   conclusions 11
   errors 11
   ethics 13
   hazards 7
exponentials 24

F

fats 53, 54
fatty acids 53, 54
formulae 14, 16
fractions 26
fuel 45–6
Fuller, Robert Buckminster 47
fullerenes 46, 47

G

gas chromatography (GC) 52
GC–MS (gas chromatography–mass spectrometry) 6
global warming 14, 48
glycerol 53

gold 10
graphite 46, 47
graphs 32
  continuous data 33
  discrete data 33
  line of best fit 33
  orders of reaction 35–6
  proportionality 34
  rates of reaction 34–5
  tangent 34–5

H

hazards 7
heart disease 9, 53, 54
helium 43
hydrochloric acid 22
hydrogen 46, 50
hydrogenation 50, 54
hydrolysis 33–5
hypotheses 4, 9–10
  anomalous results 4–5, 8, 10, 11
  bias 12
  causal links 9–10
  correlation 9

I

Iijima, Sumio 46
independent variable 8–9
infrared spectroscopy 50
ionisation energies (IE) 32, 41, 42
iron-60 6, 39
irritable bowel syndrome 52
isomerism 37
isotopes 15
isotopic masses 6, 15

K

Kekulé, August 50, 51
Kevlar® 55
kilopascals (kPa) 18
Kratschmer, Wolfgang 47
Kroto, Harry 47

L

Langmuir, Irving 43
lasers 47
Lewis, Gilbert 43
line of best fit 33
linoleic acid 53
logarithms 27, 29, 32

M

magnesium 15, 24, 32, 41
magnesium hydride 46
magnetite 39

mass
  isotopic 6, 15
  mass numbers 15
  molecular 6, 16
  moles 15–16, 18, 31
mass spectrometry 6, 15, 52
mass spectrum 47
mean 30
measurements 7
  accuracy 7
  dependent variable 8–9
  independent variable 8–9
  methodology 7, 8, 10
  precision 7
  significant figures 29
  values 8
menthol 52
menthone 52
metal hydrides 46
methane 36
methodology 7, 8, 10
models 4–5
molar gas volume 18
molecular formula 14, 16
molecular masses 6, 16
  relative molecular mass 16
molecules 5, 14, 43, 44–5, 47, 50
double bonds 50, 51, 53, 54
  electron pair repulsion model 44
  fullerenes 46
  representation of 36–7
moles 15–16, 18–20, 31
  aqueous solution 20–1
  Avogadro constant 16, 18
  molar gas volume 18–19

N

Nature 43, 47
neon 15
nitric acid 49
nitrogen oxides 14, 49
noble gases 6, 43
Nomex ® 55
nuclear fusion 39
nucleus 40, 41
nylon 48, 55

O

observations 4
oils 53, 54
oleic acid 53
olive oil 53
order of magnitude 28, 32

oxygen 14, 19
ozone layer 9–10

P

p-orbitals 50
peer reviews 12
peppermint 52
percentages 26
petrol 45
photochemical smog 56
π-bond 50
platinum 43, 48
"plum pudding" model 4
pollution 56
polyamide 55
powers of numbers 28
precision 7, 10
predictions 4
Proalcool programme 45
protein 5

R

radiation 40
radiological cancer treatments 52
rate equations 35, 56
rate of reaction 32, 35–6
ratios 14, 26
receptor sites 5, 44
red giant stars 47
relative atomic mass 15, 25
relative molecular mass 16
reproducibility of data 10, 11
rhodium 48
risk management 7
Rutherford, Ernest 10, 40

S

saturated fats 53
significant figures 29–30
  adding and subtracting 29
  multiplying and dividing 29
  rounding 29
skills
  amounts 16
  anomalous data, identification of 10
  aqueous solution 14, 20–1
  calculations 27–9
  communicating scientific information 12–13
  conclusions 12
  data, analysis and interpretation of 9–10
  decision-making 13–14
  equations 14

equations, amounts and masses 17
equations, subjects of 31–2
equations, volumes of gases 19–20
error, assessment of 11
ethics 13
fractions 26–7
graphs 32–6
logarithms 27, 29, 32
means 30
measurements 7
methodology, data and evidence, evaluation of 10
molecules, representation of 36–7
moles 16, 18–19
observations, explanation of 4–5
observations, validity of 6
orders of reaction 35–6
percentages 26–7
rates of reaction 34–5
ratios 26–7
relative atomic mass 15
relative molecular mass 16
results, recording of 8–9

risk, management of 7
significant figures 29–30
standard form 16, 24–5
techniques 5–6
titration 21–2
understanding how science advances 5
volumes of gas 18–19
writing 12–13, 22–4
sodium chloride 14
sodium hydroxide 20, 21, 22
solute 20
solution 20
solvent 20
spectrometry 6, 15, 52
standard form 24–6
adding and subtracting 25
exponentials 24
multiplying and dividing 26
state symbols 14
stearic acid 53
subatomic particles 40
sugar cane 45
sulfur 26
supernovae 39
synthesis of organic compounds 5

T
tangents 34–5
theoretical yield 17
theories 4–5
thermometers 7, 8
Thomson, J. J. 40
titration 8, 9, 21–2, 30
TNT (trinitrotoluene) 51
trans fatty acids (E-fatty acids) 54

V
values 8
vegetable oils 53, 54
volume
moles 18–19
volumetric analysis (titration) 21–2, 30

W
writing 12–13, 22–4
clarity 23
specialist vocabulary 24
standard form 24

X
X-ray crystallography 44, 50
xenon hexafluoroplatinate 43